T0039086

Alien
Earths

Alien Earths

+ ✦ +

The New Science of Planet Hunting in the Cosmos

Lisa Kaltenegger

ST. MARTIN'S PRESS 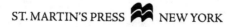 NEW YORK

First published in the United States by St. Martin's Press, an imprint of St. Martin's Publishing Group

www.stmartins.com

Art by Peyton Stark

The Library of Congress Cataloging-in-Publication Data is available upon request.

ISBN 978-1-250-28363-4 (hardcover)
ISBN 978-1-250-28364-1 (ebook)

First Edition: 2024

1   3   5   7   9   10   8   6   4   2

To Lara Sky, who makes every day a beautiful adventure;
to friends and family all over the globe, who make our
Pale Blue Dot such a wonderful world;
and for
everyone who ever looked up at the sky and
wondered if we are alone

CONTENTS

Alien
Earths

Introduction: A Message from Our Pale Blue Dot

Red puffy clouds fill an orange sky, high above the purple moss that dots the few exposed islands on the horizon. Waves break on the small stretches of shore, glittering in the red light from the sun overhead. You wait for the sunset and the darkness of night, but they never come. To experience nightfall, you have to travel for days to the far side of this distant planet, a place of eternal dusk. Even farther on, the dim light recedes, turning the landscape into a never-ending night.

The conditions greeting you on the night side of the planet are staggeringly different. The beam from your flashlight only penetrates the pitch-black of your immediate surroundings. It allows just a narrow glimpse into an unfamiliar world populated by life-forms you've never before encountered. In the gloom, you can just make out tiny bright specks of light, a biofluorescent glow that paints a slightly green sheen on the eerily alien landscape.

But the organisms here are perfectly adapted to the perpetual night. Having always lived in total darkness, they do not

require the sun's light for energy or to assess their environment.
By sensing heat and sound, they perceive the world as clearly
as humans do with sight. Like creatures in the deepest, darkest
parts of Earth's oceans, they are strangely familiar and yet not
at all.

Are we alone in the cosmos? The question should have
an obvious answer: yes or no. But once you try to find life
somewhere else, you realize it is not so straightforward.
Welcome to the world of science, which always starts with
a (deceptively simple) question.

We live in an incredible epoch of exploration. We are
discovering not merely new continents, like the explorers
of old, but whole new worlds circling other stars. Since
the first *extrasolar* planet was discovered in 1995, astronomers have found more than five thousand others in our
cosmic neighborhood. Astonishingly, that means about one
new world discovered for *every day* since we built the first
instrument sensitive enough to detect them. And we have
only spotted the ones that are easy to find—the tip of the
iceberg.

Planets are so common that they circle most stars. And
our galaxy, the Milky Way, hosts about two hundred billion stars. This staggering number indicates that there are
billions and billions of new worlds to explore in our galaxy
alone. The imaginary planet I described above could be one
of them, half dipped in constant sunlight and half in never-
ending darkness.

We are not discovering these new worlds in ships, not

even ones designed to sail through space, because these exoplanets are trillions of miles away. Those vast distances make the search much harder. But light and matter interact, which gives us a way to explore these new worlds on our cosmic shore—even though we cannot reach them yet. Just as the stamps in a passport tell you what countries a traveler has visited, light contains information about where it has been on its journey. Signs of life are written in a planet's light—if you know how to read it.

Look up at the sky tonight and count the stars that you see. For thousands of years, humans have scanned the heavens and wondered if we are alone in the cosmos, but with limited means to explore an answer. What has changed is that now we know most of these stars have companions—they harbor planets too dim for you to spot. Could there be someone watching our Earth right now, also wondering if they are alone or not? For the very first time, we have the technology to investigate.

What should we be looking for in our search for extraterrestrial life? One astronomer half-joked that we might look for large groups of animals, like pink flamingos, on other planets, though they'd have to stand still long enough for us to spot them. Color is an important tool in our search for life, but luckily, searching for colorful flamingos on other planets is not our only option. Looking a little closer we find that our planet harbors an astonishing diversity of life that changes our air and the color of our world, from bone-dry deserts to the frozen ice fields of glaciers to the hot sulfur springs of Yellowstone National Park.

Although the forms of alien life will likely differ, these organisms provide clues for our search: a mixture of the familiar rules of physics and the laws of evolution should produce organisms that could be entirely unlike those we recognize but perfectly adapted to their worlds.

Today, solving the puzzle of these new worlds requires using a wide range of tools like cultivating colorful biota in our biology lab, melting and tracing the glow from tiny lava worlds in our geology lab, developing strings of code on my computer, and reaching back into the long history of Earth's evolution for clues on what to search for. With Earth as our laboratory, we can test new ideas and counter challenges with data, inspired curiosity, and vision. This interaction between radiant photons, swirling gas, clouds, and dynamic surfaces driven by the strings of code within my computer creates a symphony of possible worlds— some vibrant with a vast diversity of life, others desolate and barren.

●

I spend my days trying to figure out how to find life on alien worlds, working with teams of tenacious scientists who, with much creativity and enthusiasm and, often, little sleep but lots of coffee, are building the uniquely specialized tool-kit for our search. I never thought that I would be part of one of humankind's most exciting adventures: searching for life in the cosmos. My curiosity about our species' place in the universe has led me from Austria to Spain, to the Netherlands, the United States, and Germany, and then back to the U.S. to head a team of incredible thinkers trying to do just that.

In *Alien Earths*, I will take you on an exciting and surprising journey as we search for life in the cosmos. I'll provide an insider's guide to what scientists are learning from Earth's history and its astonishing biosphere; I'll describe about a dozen of the most unusual exoplanets we've found; and I'll explain how these discoveries shed light on one of the most enduring questions in all of science: Are we alone?

What we have discovered about some of these new planets has been completely unexpected—some are covered with oceans of magma while others are scorched, puffy balls of gas, whizzing close to their parent stars. These fascinating new planets have shaken our worldview. Still, some of them are starting to look just a little bit like home.

So far, despite wild claims to the contrary, we have not found any definitive proof of life on other planets. Until we do, we will continue to improve our toolkit and look for signs of alien life the hard way: searching planet by planet and moon by moon.

The most exciting phase is about to begin.

OUR GALAXY: THE MILKY WAY

|———————————————————————————————|
100,000 LIGHTYEARS (TRAVEL TIME)

SOLAR SYSTEM
(25,000 LIGHT-YEARS FROM CENTER)

TRAVEL TIME FOR LIGHT FROM SUN

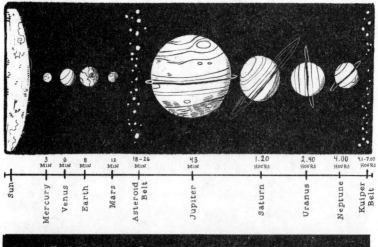

	3 MIN	6 MIN	8 MIN	12 MIN	18–26 MIN	43 MIN	1.20 HOURS	2.40 HOURS	4.00 HOURS	4.1–7.00 HOURS
Sun	Mercury	Venus	Earth	Mars	Asteroid Belt	Jupiter	Saturn	Uranus	Neptune	Kuiper Belt

LIGHT SPEED ~ 6 TRILLION MILES PER YEAR

1 sec: Earth – Moon (240,000 miles)

8 min: Earth – Sun (93 million miles)

10 to 600 days: Sun – OORT Cloud
(0.2 – 9 trillion miles)

4 LY: Sun – Proxima Centauri
(25 trillion miles)

Chapter 1

At the Brink of Finding Life in the Cosmos

These worlds in space are as countless as all the grains of sand on all the beaches of the Earth. Each of those worlds is as real as ours and every one of them is a succession of incidents, events, occurrences which influence its future. Countless worlds, numberless moments, an immensity of space and time.

—Carl Sagan, *Cosmos*

The First Images from a New Spacecraft

The foam on my Portuguese espresso tastes a little bitter, but I hardly notice. For the past hour I've been staring at the images on my computer screen, a live feed from NASA's recently launched James Webb Space Telescope (JWST). The screen is dark now, and my own thoughts are wandering into that darkness, into what mysteries of the cosmos it will reveal.

Watching the flawless launch of the JWST on a late-December day in 2021, scientists from every continent focused intently on each step of the launch. From liftoff to observations, JWST had 344 single points of potential failure—each one of which alone could take down the entire system—so although we were relieved every time something went right, we knew that there were still hundreds of things that could go wrong.

With my eyes glued to NASA TV (and to our team's Slack channel, where colleagues in every time zone commented on the success of each step), I kept reminding myself to breathe—in and out, in and out. There was nothing any of us could do because the rocket had lifted off and was on its way to its final destination in space, a place about one million miles away called the second Lagrange point, L2. *New York Times* reporters Dennis Overbye and Joey Roulette described the image of JWST blasting off into space beautifully, as looking like "a tightly wrapped package of mirrors, wires, motors, cables, latches and willowy sheets of thin plastic on a pillar of smoke and fire." It also carried the dreams of thousands of scientists like me, hoping to catch a glimpse of the cosmos that had been beyond our reach—and our view—until now.

The JWST is the first telescope capable of capturing just enough light with its 21.3-foot (6.5-meter) mirror to explore the chemical composition of the atmosphere of other rocky worlds. Size is the key to collecting light. Imagine a bucket—the larger it is, the more rainwater it can catch in a downpour. The telescope's mirror operates the same way—the larger it is, the more light it can collect.

The cheering of the crew in the control room at the successful separation of the telescope from the rocket interrupted my thoughts. The final image from the launch broadcast was a close-up of the telescope drifting into the darkness of space, with a mesmerizing view of Earth's blue globe in the upper corner.

It would take months before the JWST successfully navigated each of the remaining hurdles as this beautiful bundle of mirrors, cables, and solar panels unfurled. After that it had to slowly cool to the freezing temperatures at which it could begin to operate.

It was only when examining the first signals that we proved we had beaten the odds. Having cleared every point of potential failure, this amazing telescope operated flawlessly, providing the first glimpses of a new way to see our cosmos—and a taste of the astonishing discoveries yet to come.

One of the most stunning images captured by the JWST is of the Eta Carinae Nebula, about seven thousand light-years away. This stellar nursery, where stars and new planets are just forming, looks like celestial art painted by a cosmic brush. But it's not just the birth of new worlds that the JWST is unveiling. An image shared with the public in July 2022 by President Joe Biden, a day before the first official NASA data release, revealed a time when the universe itself was in its infancy. In the JWST deep-field image, you can see thousands of galaxies scattered like glittering dots across the black canvas of space, contained in an area of the sky that, viewed from Earth, is about the size of a grain of sand. Their light took more than thirteen billion years to reach us, sending us a message from a time long before Earth was born. On its journey to our telescope, some rays were bent as they passed a massive cluster of galaxies. Matter and light interact, so this ancient light was warped into the beautiful arcs seen in the image, revealing relativity's pull over space and time.

This view of ancient galaxies fills me with wonder—and hope. There are billions of stars in this image, ancient echoes of other possible worlds. In this tiny corner of our cosmos, planets could have been formed countless times, and yet our *now* and their *now* do not intersect because of the vastness of space between us. While some stars—like those in the deep-field image on my computer screen—are lost in time to us, a myriad of closer stars remain—with intriguing worlds circling them. And now we can explore the closest ones.

In science, by finding new ways of seeing, such as catching the light from dimmer objects in the huge mirror of the JWST, we are able to detect what we could only imagine before. New insights transform our understanding. These images are a touching testament to the cooperative spirit of humanity: it took thousands of people from all corners of the world to make that vision a reality.

Among the first images released, one revealed a detailed view of the light from a scorching, puffy giant planet, WASP-96b, shrouded by layers of clouds, haze, and steam. It blasts around its star twice a week. While devoid of life, the JWST image proved that the space telescope can explore the atmospheres of other, smaller Earth-sized planets given more time. Places where life could thrive. As a member of one of the scientific teams behind the JWST, I work with a creative group of scientists to explore these new worlds on our cosmic horizon.

Discovering life on another planet would forever revolutionize our entire worldview.

So Where Is Everyone?

Let's assume, for a moment, that the universe is teeming with life. In that case, the obvious question is: Where is everyone? In my introductory astronomy class, "From Black Holes to Undiscovered Worlds," I ask my students to suggest possible explanations for why we have had no credible record of alien visitors to date. I am going to skip all discussion of supposed UFO sightings here, since the subject is full of poor observations and would require a book-length reply like Carl Sagan's thought-provoking *The Demon-Haunted World*, one of my favorite reads. Among many other insightful points, Sagan asks why alien species that have surpassed us so monumentally in technology that they can travel from star to star would need to abduct a whole person to study. Even a comparatively less advanced species like us humans has developed the technology to take DNA samples from hair or saliva. Wouldn't collecting these samples from unsuspecting humans be a much more effective way to study them than beaming people up into their spaceships one by one? For the record, most of my students' theories involve either doomsday scenarios—alien civilizations have destroyed themselves before they can reach out to find others—or the endless void—we haven't seen anyone else because we are the only life the cosmos ever produced.

This puzzle of absent aliens is not new. Enrico Fermi, the Italian-American physicist and Nobel Prize winner, famously posed the question "Where is everybody?" in a conversation on the possibility of extraterrestrial life in 1950. If technological civilizations are common in the universe, surely some would have developed sufficiently to visit or at least contact us by now? This mystery is known as the *Fermi Paradox*; the discrepancy between the absence of evidence for advanced extraterrestrial life and the high likelihood that it should exist. Casting a dark shadow over the conversation at the time was the fact that scientists were in the midst of developing nuclear weapons that could wipe out civilization on Earth.

How abundant might intelligent civilizations be within our vast universe? One way of thinking about this was proposed by the American astronomer Frank Drake, a pioneer in the search for extraterrestrial intelligence (SETI), who in the 1960s developed a systematic process to assess SETI's prospects. Searching for what he called "a whisper we can't quite hear," he wove together a number of factors in a framework that is called the *Drake Equation*. The seven interlinked factors began with well-constrained estimates of the rate of star formation, educated guesses on the likelihood of which ones might have planets circling around them, and the fraction of these capable of nurturing life, before continuing on to wild speculations about the probability of life actually evolving, to the fraction of life-forms that might develop intelligence, and the even smaller percentage that might be capable of interstellar communication. The very last factor of the Drake Equation poses a question that can evoke either boundless enthusiasm or chilling pessimism about our chances to connect with any alien civilization: How long can technological civilizations survive?

The vastness of space is only occasionally dotted with stars, with enormous distances between them. For me, these become much easier to imagine when I shrink them in my mind to the scale of everyday objects. Let's reduce our solar system—from the Sun to the outermost planet, Neptune—to the size of a cookie with a diameter of about two inches. How far away do you think the Sun's closest neighbor is? Two cookies away? Five? One hundred? It is much farther than that—nearly nine thousand cookies. Or, on the same cookie scale, about four football fields away. To chart the distances between stars in the cosmos, you need larger units than miles or kilometers, or cookies. Using a light-year as our cosmic yardstick makes it easier to comprehend these unimaginable expanses.

Light travels at an incredible speed: about 190,000 miles (~ 300,000 km) per second, or an astonishing 6 trillion miles (~ 9 trillion km) per year. It takes light only about one second to travel between the Earth and the Moon, about 240,000 miles (~ 380,000 km), and a mere eight minutes to cross the distance from the Earth to the Sun. In those eight minutes, light travels a relatively tiny cosmic distance of 93 million miles (~ 150 million km). The closest neighboring star to our Sun is Proxima Centauri, at an enormous distance of 25 trillion miles (~ 40 trillion km) away. Even light takes about four years to travel that vast distance. So as well as conveying distance, a light-year scale tells us how long it takes for light to make the trip. We humans are starting to venture into our solar system, but these distances are small compared to the spaces separating the stars.

Our galaxy is about one hundred thousand light-years across. If a civilization had the means to navigate at even 10 percent of the speed of light, it could, in principle, cross the galaxy in about a million years. In principle. Most of the travel time would be spent voyaging through empty space: even a trip between our Sun and its closest stellar neighbor would take decades. Most of the trip would be endlessly boring because the distances between stars are so vast. And moving at such breakneck tempo would be exceedingly dangerous, since, running into even a small grain of interstellar material at that speed could result in disaster for the spacecraft and everyone on it. A million years is a long time compared to a human lifetime, or even to humanity's evolution, but some stars and their planets are much older than ours. If older civilizations exist, our galaxy might already contain their outposts, relics, or signals indicating advanced technology. But we have not encountered any yet. (Nor have we traveled very far from home.) So, as my students often suggest, do we lack alien visitors because the distances between habitable worlds are just too vast to navigate?

Let's leave reality and get inspired by solutions presented by science fiction. While I personally love the idea of moving at faster-than-light speed, like the fictional starship *Enterprise* in the sci-fi *Star Trek* franchise, warp speed is most likely impossible to achieve, even in the future, because our universe is bound by the laws of physics. Based on everything we know, faster-than-light travel is a barrier we cannot cross. In an alternative visually stunning vision imagined in Luc Besson's 2017 film *Valerian and the City of a Thousand Planets*, through complex but possible marvels of technology, enormous space stations navigate the cosmos while their passengers experience the wonders of the universe. The fictional spaceship contains a vast metropolis that is home to species from a myriad of alien worlds.

For now these possibilities to cross the galaxy are beyond us. But perhaps there is another way aliens might reach us. Since light travels at an astonishing speed, that means messages encoded in radio signal can travel fast. Often, the word "light" is used only to describe the narrow range of electromagnetic radiation that our eyes have evolved to see. Imagine that you are holding a prism, a wedge of glass, and you pass a beam of white sunlight through it. A brilliant cascade of colors emerges ranging from deep reds to vibrant violets: the spectrum of visible light. Yet, what you see is but a minute portion of the full range of electromagnetic radiation that extends far beyond our human sight, into the infrared and ultraviolet, radio waves and gamma rays, all different notes in this grand cosmic composition of light.

One way to find advanced, communicating civilizations would be to collect radio signals being beamed our way that are not naturally occurring. While astronomical objects like galaxies generate radio signals as well, scientists are looking for signals that stand out, maybe—a kind of cosmic greeting. But these interstellar greetings would dissipate in the vastness of space. Every doubling in distance reduces the signal strength to one-quarter of its previous volume, so at a certain distance, even the loudest shout becomes an imperceptible whisper—and that is assuming anyone is listening. Astronomers are looking for these radio signals but they have not found any yet. Does that really mean that there is no other life in the cosmos?

The Great Silence

Giant traveling space stations are not yet available, and we can't break the known laws of physics, so this Great Silence of the cosmos looms dauntingly. This has led scientists (and my students) to suggest the possibility that even if life had existed somewhere else in the past, some barrier like a cataclysmic event has destroyed it and prevented civilizations from venturing into our galaxy—a Great Filter, so to speak, that has so far prohibited alien intelligence spreading through the cosmos. This Great Filter could lie in our past. For instance, maybe it is astonishingly complicated to start life on a planet. Or what if it's easy for life to begin but almost impossible for it to get past the earliest microbe stage? If alien life did become intelligent and technologically savvy enough to build satellites and capable of sending spaceships traveling through a planetary system, that technology might also be powerful enough to destroy every corner of their planet. Or the cataclysmic filter could lie in our future. How hard is it for a civilization to survive its own technological growth? Maybe other life-forms have destroyed themselves before they could travel to the stars. A very depressing thought. But on the bright side, in that scenario, they are a much bigger danger to themselves than to us. Nuclear bombs and climate change are just two of many possibilities that could lead to the destruction of a civilization.

But why do we automatically assume that other civilizations would even want to visit or communicate with us? Let's set aside the issue of what atmosphere and environment potential alien visitors would need to survive; how intriguing would Earth appear as a destination?

Imagine that you could visit one of two planets: the first is five thousand years younger than Earth, and the second is five thousand years older. Both show signs of life and are at a similar distance. Which one would you pick? Whenever I ask this question, most people pick the older, more advanced planet. Let's assume a fictional alien civilization was given the same choice. Using that reasoning, our spectacular planet becomes a bit less interesting. Don't get me wrong, Earth is my favorite planet, but in terms of technology, we are just getting started. True, twelve astronauts have visited the lunar surface, but so far, human beings have not even reached the nearest planet, let alone the nearest neighboring star. Given a choice, would Earth really be the planet to pick—yet? In the optimistic case of a cosmos teeming with friendly worlds, the Earth is not yet at the grown-ups' table.

The premise that anyone who could call us would do so immediately, seems flawed, making the Great Silence less eerie.

Talking to a Jellyfish

If we ever found another civilization that we could communicate with using radio signals or visible light—both of which travel much faster than spacecraft—I often wonder what we would say. What questions would we ask? And how would we ask them? It seems unlikely that they would understand English, Chinese, Spanish, or any of the other thousands of languages spoken on our beautiful planet. The experience might end up being like a human trying to talk to a jellyfish. I've attempted that; the results were less than promising. And in that case, the jellyfish was right there in front of me. I could see it and could have touched it (but I refrained), and I listened for any sounds it might make in an attempt to learn its language (with no success). Note that I am not an expert at interspecies information exchange, though there are other scientists worldwide studying the communication of dolphins, whales, chimpanzees, and dogs, among others—they might fare better. To interpret and understand other species, it is critical to observe actions and other visual cues and combine these with your interpretation of sounds. It is a daunting task. Imagine how much harder it would be with a civilization you couldn't even see? An advanced interstellar civilization trying to converse with a less advanced one would be a bit like humans trying to interpret the movement of a school of fish, which is dynamic, purposeful, and even beautiful but, ultimately, puzzling in its intentions.

Even though humans are only at the beginning stages of our attempts to communicate with other species, spacefaring civilizations should have something in common with humans. To find other civilizations and correspond over cosmic distances, they'd need to understand how the cosmos works. And while humans have employed many methods to do that—from reading tea leaves to random guessing—there is only one accurate way to figure out how planets move and how spacecraft and radio signals operate: the scientific method. The scientific method is brutal in that it does not care what you hope to find, but that is also its greatest strength: with new facts, new ideas emerge and replace outdated notions. It forces you to find reliable information—a key any species would need in order to discover new planets and to send or search for messages, let alone invent safe means of space travel to get there.

Here Be Bananas, Aliens, and Dragons

I once began a lecture in my introductory class by holding up a banana and asking my students, "Could this banana be an alien?" Let me be clear: I don't think a banana is an alien—or at least I think it is extremely, *extremely* unlikely. But a banana was the only unusual object I could find in my backpack, and I wanted to make a point. How do we really know if something is an alien or not?

To find life in the cosmos, we need to stretch our minds and search at the limits of technology. Not only do we need to work at the edge of knowledge, but we must ask the right questions and overcome our own biases. The human brain has evolved to spot patterns—a great evolutionary trait for people who were once hunted as prey. If your ancestors spied hungry lions hidden in tall grass before the lions sneaked up on them, they survived. If there were a few false alarms and a bit of energy wasted in fleeing unnecessarily, that was not as bad as being surprised by lions on the prowl. So our ancestors learned to recognize the presence of predators by the smallest changes in the environment—bending grass, a sudden eerie quiet, or slight movement in the bush. Many tiny signals together could alert them to danger. That ability to discern patterns is still useful, but it can also make us think we see things that are not actually there.

Take, for example, the human face many people thought they recognized in some old NASA images of a rock formation in the Cydonia region of Mars. This led to endless wondering about whether aliens had left us a message inscribed in the Martian landscape. But isn't it curious that it was a human face and not, say, the face of a dog or a panda? Perhaps that revealed an unconscious hope that aliens were just like us. Clearer photos later showed that these Cydonia rocks could be mistaken for a face-like image only at low resolution and when the sunlight hit them just right. But the episode serves as a helpful reminder that our species' ability to see patterns can be misleading when we try to make sense of new information. One of the advantages of the scientific method—or disadvantages, depending on whom you ask—is that it requires you to accept what the nineteenth-century British biologist Thomas Henry Huxley called "the great tragedy of science—the slaying of a beautiful hypothesis by an ugly fact."

Asking the right questions helps us determine what is a real pattern and what is just random noise. Let's return to my banana and start asking questions. What is the banana made of? Where did it come from? Does it resemble other items we are familiar with? Does it share chemical or genetic properties with other recognizable Earth objects? Does it behave in a novel way? As it turns out, we know from hundreds of years of agriculture where bananas grow, we know that they have grown on Earth for a long time, and we know how they evolved on our planet. So we can be pretty sure that bananas are not aliens, and we can use the same thought process to determine that neither you, me, nor your coffee cup is an alien. However, other claims are not so easily dismantled.

Let's conduct a more challenging thought experiment: In another lecture, I offer my class the opportunity to buy a dragon. It will be a good investment, I announce—who wouldn't want to own a dragon? At first, I get a lot of interested potential buyers. Then I ask for fifty thousand dollars, and the questions start. Can my students see it first? The answer is no, because my dragon is invisible. Can they touch it? The answer is no to that too. Can they hear it roar or breathe fire? No, because this particular type of dragon is silent and does not breathe fire. When I ask for offers again, I notice that the prospective buyers' initial enthusiasm has evaporated.

Unfortunately, I don't have a dragon to sell, but the example encourages my students to use the scientific method to avoid getting scammed. Given the hypothesis that a dragon exists, they devise tests to prove it. If all the tests you devise fail, then no dragon exists, at least as far as you can tell. No one would give me fifty thousand dollars for a dragon no one can see, hear, or touch. Slyly, the scientific method had taken over their thinking.

People may automatically apply the scientific method to their dragon-buying decisions, but curiously, they don't do it as often with other remarkable claims. Let's say someone promises that if you give him fifty thousand dollars, he will show you proof of alien life; again, that is really a steal if it turns out to be true. When I ask my students what proof would convince them to pay up, a lively debate ensues. What if you can't see, touch, or hear it? What if the proof is just a smudge on a photo this person shows you? Is alien life the only explanation? Finding the first sign of alien life is an incredibly tempting prize. But this is where the scientific method uncovers impostors: if only one person claims it, be wary. Other scientists must be able to independently confirm the results and observations, and so far—unfortunately!—for every initial alien sighting or discovery, we have not found any proof that holds up under further investigation. "Extraordinary claims require extraordinary evidence," Carl Sagan once wrote. Proof of alien life needs to hold up under intense scrutiny because it would indeed be extraordinary.

Another key point I emphasize to my students is that you can solve a problem only if you can describe it. And to do that, you need to find the right language. The language that reveals the mysteries of the cosmos is mathematics. The advantage of that language is that it is the same wherever you go. Once you learn it, you can speak to other scientists across the globe, creating a vast interconnected network of thought. With this language, I can use digital code to "paint" imagined worlds on my computer screen. My canvas is my laptop—new planets arise from strings of digits encoded in an extensive computer program that captures characteristics like heat, moisture, and gravity to model planets that circle other stars. My ultimate goal: I want to know if these new worlds could support life and how to find it.

With such tools, there are ways to overcome the handicap of not having interstellar starships to scout for life in the cosmos. Any widespread biosphere on a planet will most likely change that planet: it did on Earth. As an example, about two billion years ago early life-forms on our planet created so much oxygen as waste that the atmosphere was transformed. Such events give us a way to locate the presence of life in the cosmos, whether it wants to communicate with us or not. If organisms did not impact the biosphere on their planets, our search there would be in vain, and we could only wait and hope that someone eventually wanted to send us an interstellar message.

One way to find out if a spacecraft could spot signs of life on an inhabited world—without relying on messages—is to analyze our own Pale Blue Dot from space. The Galileo spacecraft, the first mission to orbit Jupiter and release a probe to penetrate the giant planet's atmosphere in 1995, used an earlier flyby of Earth a year after its launch in 1989 to pick up speed for its journey—during which it inspected our planet. Carl Sagan used that information to decipher what Earth's signatures of life looked like from space. This was a first test to prepare for the future, when a telescope would be capable of catching the light from such a world circling another star. Seen from space, Earth presents a combination of gases that scientists can only explain by the existence of life. While detecting the presence of gases might not have the same impact as a message written in the stars saying *Hello, Earthlings*—or the equivalent in jellyfish language—it provides an independent chance to search the cosmos for other life-forms, whether they want to communicate or not.

The Golden Record: A Message in a Bottle

When Voyager 1 and Voyager 2 were launched in 1977 to explore the outer planets of our solar system, NASA included a message from humankind on each spacecraft: the Golden Record. Inscribed with the words "To the makers of music—all worlds, all times," it is a time capsule of life on Earth. Carl Sagan led the team that compiled the Voyagers' interstellar message, a phonographic survey of our technical and artistic accomplishments, with the inspiring Ann Druyan, who served as its creative director, and would later become a Peabody and Emmy Award–winning writer, director, producer, and Carl Sagan's wife and collaborator.

The Golden Record tells the story of our planet: a story captured in images, sound, and science: 115 images of life on Earth and 90 minutes of musical selections from different cultures and eras; natural sounds made by surf, thunder, and wind; songs by whales and birds; human sounds like laughter; and spoken greetings in 55 languages, including a "Hello from the children of planet Earth." (See the Golden Record Playlist at the end of the book.)

Why a record? Because you can describe in simple terms how to play a record to someone who has never seen one. Both Voyager probes carry a stylus aboard so another civilization could make its own record player. The cover on the record instructs the recipient to place the stylus on the outer edge of the record to play it. The golden disk should be played at 3.6 seconds per rotation. But an Earth second is a somewhat arbitrary time interval that is based on the 24-hour rotation period of modern Earth; even on other planets in our solar system, an Earth second carries no meaning. None of them rotates every 24 hours. So other civilizations would likely define completely different time intervals. How can you translate the speed at which a record should be played into a cosmic standard time? The team solved this problem by using a time constant that any spacefaring civilization should understand, a fundamental characteristic of a hydrogen atom: the time it needs to transition between the lowest and the second-lowest energy state—roughly 0.70 billionths of a second. About five billion of them translate into the 3.6 Earth seconds.

The records are etched in copper, plated with gold, and sealed in aluminum cases, providing a chance to still read them after more than a billion years. Gifts to the cosmos from one Pale Blue Dot. Can alien civilizations build record players? Maybe, maybe not. But the key is to show that the record contains information. It is a message created without knowing how another life-form might experience the world. Organisms will interact with their environment, and even if these beings just feel the structure of the disk, humankind's cosmic message can be deciphered.

The record cover also features a map, pinpointing the location of our solar system with respect to pulsars. A pulsar is a collapsed stellar core left over from a gigantic explosion at the end of a massive star's life; as we'll explore them later. Pulsars can be seen over large cosmic distances beyond galaxies, and they can be identified individually by how many signals per second each one sends. The Golden Record team put our solar system on the cosmic map by indicating its location in relation to fourteen nearby pulsars.

Also aboard is a source of uranium-238, which acts as a clock. Uranium is naturally radioactive, which means its nucleus is unstable, and it steadily decays into its daughter elements. Half of the uranium-238 will decay in four and a half billion years, so measuring how much uranium-238 remains versus how much there is of its daughter elements tells the recipients of the records how long the probes have been traveling and when they were launched. They started their cosmic journey the year I was born. Only the tiniest bit of uranium-238 has decayed since then. The Golden Records will fly through space for billions of years to come, mounted on the outside of the Voyager spacecraft.

In the vastness of the cosmos, the probability that the probes traveling through the enormous expanse between the stars will be found is minuscule. Both Voyager probes were flung out of our solar system after they finished their primary mission to study our gas giants. With what little power they have left, they are analyzing interstellar space to give scientists the first insights into what happens when our Sun's influence wanes. Only advanced spacefaring civilizations could locate the spacecraft in the cosmos because

Voyager 1 and Voyager 2 were not designed to land any-where. They are not traveling to any specific destination. It will take about forty thousand years for the spacecraft to pass anywhere near another star. And they will only get within about ten trillion miles of the star Gliese 445 for Voyager 1, and Ross 248 for Voyager 2. Both are cool red stars in our cosmic neighborhood.

If a Golden Record is found millions or billions of years from now, it might be the last remnant of our civilization. It might even be our most enduring achievement. It captures us at a moment in our evolution when we could send a physical message into the cosmos. In the 1978 book *Murmurs of Earth* by Carl Sagan and the Golden Record's team, Ann Druyan wrote, "Voyager moves among the stars, bearing its cargo of echoes and images, and, in the logic of such distances, it keeps us alive."

What would *you* put on a record to communicate what it means to be human using a mere 115 images and 90 minutes of music? Trying to put such a set list together in my mind makes me realize what a breathtaking journey humankind is on. The Voyager missions are still journeying toward the next stars. And while we don't know if any aliens have yet listened to the Golden Records, Carl Sagan eloquently summed up their significance: "The spacecraft will be encountered and the record played only if there are advanced space-faring civilizations in interstellar space, but the launching of this 'bottle' into the cosmic 'ocean' says something very hopeful about life on this planet."

One song on the record always touches me deeply: "Dark Was the Night, Cold Was the Ground," recorded in 1927 by Blind Willie Johnson, a Texas blues musician. In 1945, Willie Johnson's home was destroyed by a fire, but he lived in the ruins because he had nowhere else to go. He contracted malaria but hospitals refused to treat him, either because he was Black or because he was blind—the accounts differ. We don't even know where he is buried—but his song is aboard two spacecraft en route to the stars. And maybe, someday, someone will find one; a beacon of hope in the darkness.

A Pale Blue Dot

Before Voyager 1 got the final push to leave our solar system, Sagan convinced NASA to turn the spacecraft around and take a last image of the Earth, its home planet. This spectacular photograph captured on Valentine's Day in 1990, more than three decades ago, shows Earth as a tiny point of light suspended in a sunbeam on the dark canvas of space. The vast oceans and a medley of clouds combine to paint it pale blue. That picture changed how I think about our planet.

The image of our tiny Pale Blue Dot reminds me every day of just how beautiful and, at the same time, how fragile our world is. All that protects us is a thin sliver of the atmosphere. Most of our air is contained in the first six miles (~ ten km) above the ground. If you could go on a leisurely road trip straight to space, let's say at about thirty miles (~ fifty km) an hour, it would require only about a dozen minutes to pass through that whole region. If the Earth were the size of an apple, our atmosphere would be thinner than the apple's skin. To survive, humanity has to take extremely good care of this thin layer that protects us from certain doom.

The photo was recorded from only about five and a half light-hours away (about four billion miles or six billion km from the Sun), a distance comparable to just beyond Neptune.

No other spacecraft has yet taken a photo of Earth from farther away, capturing an image of our own world "as a mote of dust suspended in a sunbeam," as Sagan wrote in his beautiful description of our planet in his book *Pale Blue Dot*. The photo of Earth was taken just thirty-four minutes before Voyager 1 powered off its cameras forever: the last glance back toward home for the final time.

Chapter 2

How to Build a Habitable World

I look at the natural geological record, as a history of the world imperfectly kept, and written in changing dialect; of this history we possess the last volume alone, relating only to two or three countries. Of this volume, only here and there a short chapter has been preserved; and of each page, only here and there a few lines.

—Charles Darwin,
On the Origin of Species (1859)

Past, Present, Future

My world initially consisted of our house in the Austrian countryside with its enchanted garden and the old, giant trees swaying in the wind on stormy days, and the few houses nearby, including my grandparents' farm. Every summer, my grandfather went out with his scythe and harvested the grass by hand to make hay. My mom and dad helped while my sister and I watched from our window as the scenery changed from fields full of long grass to rows of hay drying in the hot summer sun, altering the colors to reflect the transition in the world around me, my first introduction to the color cycle of seasons.

My best friend's house was a few minutes' walk away, and she and I went on countless imaginary quests in worlds constructed out of Lego. Our imagined excursions were anything we could dream up, and provided untold adventures. The pile of Lego pieces was one of my most prized possessions because it could become castles, spaceships, houses, or majestic mountains. In the basement I learned from my dad how to craft wooden sculptures and capture the world around me on canvas. I loved building things even then, long before I would receive my engineering degree and help design a spacecraft to search for alien Earths.

I often ventured along the fields to my grandparents' farm to feed the dozens of clucking chickens. The smell of fresh, sugary yeast pastries filled with plums and jam would lure me into the old farmhouse's small, well-loved kitchen, which was heated by an old wood-burning stove. My trips were often rewarded with my grandmother's oven-warm pastries, hinting at the delights of exploration.

My elementary school was a fifteen-minute walk away, a walk I made in sunshine, rain, storms, or snow along with the few other kids who lived along the road. I experienced the powerful weather patterns on our planet firsthand. My town and its surrounding areas consisted of a few hundred people. Everyone knew where every kid belonged and ensured that we got where we were going. It made for complete freedom. These adventures laid the foundations for my love to explore that now includes worlds far beyond our own.

In my hometown, the streetlights turned off around nine p.m., sending me into velvet darkness watched over by thousands of stars, opening my world way beyond our own planet. In parts of the sky, the stars merged into a beautiful band of light—too many stars to make out individually. This happened especially on those cold winter nights when my breath condensed in the chilly air and the stars seemed to glow ever brighter.

We spent our summer holidays in an even smaller village, in a wooden hut on the slope of a mountain in southern Austria that provided me with new views of the stars and new mysteries to ponder. The early-morning chill of late-summer days found us huddling in front of the hut's small iron stove, waiting for the heat to take our visible breath away. The water to wash and drink was freezing because we'd collect it from a nearby well, but it was a welcome icy refreshment, especially after hikes to the mountaintops, places where gray stone replaced lush vegetation, creating a completely different landscape. It is an ecosystem with its own plentiful life that was just a bit different than what I was used to seeing in our town, making me wonder what life could be like in places I had not encountered yet.

At home or on vacation, my mom was always there to listen to my stories. She also nurtured my growing love for books. When I was ten, the library gave up trying to put limits on the number of books I could borrow, and I happily carried increasingly larger piles home, sinking into worlds I could only imagine in those pages. Who could have dreamed that one day I would be the one filling a spot on the library shelf?

When I got older, my world expanded beyond Austria. On family road trips to Italy, I met people who smiled a lot but who did not understand what I was saying, teaching me how difficult it could be to communicate, even with people who were born only a few hours' drive away. My dad sometimes brought visitors from different countries home for dinner, since he worked as a civil engineer, building complex structures worldwide. I remember the Japanese business partners who bowed in greeting instead of shaking hands. I was most fascinated by their language, which sounded like nothing I had heard before. At dinner, the conversation shifted into broken English. I knew only a few English words from school, but these few words allowed me to connect. Like on my trip to Italy, knowing just a few words gave me a key to understanding something new. The visitors showed me their writing, Japanese Kanji, beautiful images formed with pen strokes that merged into a picture that created a word, revealing an alternative, novel way of communicating. That curiosity sparked my love for languages and travel—exploring our globe and learning languages in the places I lived (Spain, Portugal, The Netherlands, U.S.). It still fills me with excitement to watch words and symbols that initially made no sense transform into bridges that allow me to communicate with people all over the world.

My world began as a small town, but it grew to encompass the whole globe, then reached into the cosmos, with new planets to explore in whatever way possible. The mysterious twinkling dots of light of my childhood have transformed into scorching balls of gas and the heavens into a celestial history book of the cosmos, but when I look at the night sky I still feel the wonder and excitement I felt as a child to uncover what is out there.

Key Ingredients for a Habitable World

All around you is a vast, rippling ocean, its surface only broken by the distant peaks of active volcanoes spewing out lava and gas. For miles and miles, there is nothing but water. The nearly endless sea stretches out to the horizon, topped with ripples that carry their salty scent to the shore. Only the sounds of the waves breaking on land and the winds blowing disturb the total silence.

The islands that dot the ocean are barren. No grass, no trees, no animals anywhere. It is eerily silent. Nothing moves except for the waves and the molten rock bursting from beneath the crust. Volcanoes birth red-hot lava streams that shed their heat in the cold embrace of the vast sea.

The stars form unfamiliar patterns, and staring up at the sky, you search in vain for a recognizable constellation like Orion or the Big Dipper. A dark, giant moon looms enormous in the sky. Sunrise and sunset chase each other every few hours on this strange world.

When we look for life in the cosmos, Earth is our lone key to unlock the secrets of what it requires to get started. It all begins with a few ingredients: a rock in space, heat from its star, water, carbon, and winds on the surface of a new world.

First, you need a rock in space, like a planet or a moon. Then you need energy. The biggest energy source for our planet is sunlight. So let's add starlight to the rock in space. Because the rock circles a star, we call it a planet. If it has no atmosphere, the energy from the star will heat up the planet's surface, and the rock will radiate this energy back into space. Nothing much generally happens in this scenario, and the planet seems frozen in time, like Mercury, the closest planet to our Sun, which has looked the same for millions of years and likely will look the same for millions of years more. That is, unless something hits it, in which case it would have an additional crater.

Let's add an atmosphere. The planet only gets to hold on to its atmosphere if it is massive enough for its gravitational embrace to prevent the gas from escaping. If it is, the story of the rock in space becomes more interesting: the light from the star heats its atmosphere as well as its surface.

While a planet with an atmosphere is a much more interesting place than one without, we need another key ingredient for life as we know it: liquid water. Liquid water evaporates and turns into gas close to the Earth's surface, then rises into the atmosphere until it reaches the height where it is cold enough for it to condense again, resulting in rain or snow falling to the ground. The region around a star where a planet is neither too hot nor too cold to support oceans and seas on its surface is called the *habitable zone* or the *Goldilocks zone*, and it's the best real estate for life that can alter its world. The habitable zone around a star is a perfect place to start our search for a planet that could host life.

First: Make a Rock in Space

About four and a half billion years ago, an average—but to us very special—star and its eight planets formed. Billions of years later, life on the third rock circling this star would name that star the Sun. Four and a half billion is a staggeringly large number. To put it in perspective, 4.5 billion *seconds* is a bit more than 142 years, much longer than a human lifetime. Four and a half billion *years* ago, our corner of the cosmos was very different. In a galaxy called the Milky Way, a large cloud made mostly of hydrogen atoms and small amounts of gas, ice, and mineral grains rotated slowly within one of its spiral arms. It was piercingly cold, about −450 °F (−270 °C), very close to absolute zero. The cloud turned lazily. But then, a shock wave from a nearby exploding star triggered a catastrophic change. The cold cloud collapsed. Like countless times before in the cosmos, gravity drew the surrounding material together into a hot, dense, central mass: a young star.

Gravity, the most influential architect of the cosmos, is the attraction between different objects. Its strength increases the closer the objects are to each other and the more massive they are. It keeps us grounded, holding us on the

Earth's surface instead of letting us float into space. Gravity pulled together more and more particles in this slowly rotating cloud of gas. The more pieces stuck together, the greater gravity's power in this little part of the cloud grew and the more material it dragged toward this nucleus. Finally, its center became so hot and dense from the pressure of all the overlying material that hydrogen atoms crashed into each other with such speed that they stuck together, creating helium atoms and energy.

A helium atom is just a tiny bit lighter than the four hydrogen atoms it has been made from. And that tiny bit of difference in mass has enormous consequences. More than a century ago, the German-American physicist Albert Einstein captured the incredible power a little bit of mass that becomes energy has in his famous formula linking energy (E) to mass (m) and the speed of light (c): $E = mc^2$. The change in mass from four hydrogen atoms to one helium atom is less than 1 percent, but that tiny change is what powers our Sun, keeps Earth warm, and made us possible. The Sun loses about five million tons of mass a second. To put that in perspective, an adult blue whale weighs about 100 tons. So the Sun loses the equivalent of the mass of 50,000 adult blue whales per second.

When you look closely at our solar system, you'll notice something odd: all the planets circle the Sun in the same direction. They could orbit in any direction and still obey the law of gravity; the fact that they don't suggests that the planets formed from material that moved in the same direction around the fledgling Sun. And our planets also lie in one plane, which indicates that they all formed from the same flat rotating disk of dust and gas.

Astronomers have spotted such flat disks around other young stars. These are planetary nurseries, places where young stars are forming their own planetary systems, creating intriguing new worlds to discover. Watching them is like a glimpse into our past, when our Earth was one of the planets that had just formed out of crashing pebbles circling a nascent Sun. There is a ticking clock for planet formation. A small fraction, about 1 percent, of all the material in the cloud falling to the center did not become part of the central star because the cloud had started out already rotating. Instead, that small part bounced around until it created a flat disk that circled the budding star. In this disk, rocks, ice, and gas made of molecules such as water, carbon dioxide, methane, and ammonia smashed into each other.

First, tiny specks of rock and ice grow into pebbles, then boulders, creating ever bigger objects. With millions of those pieces circling a young star, the smallest diversions from their paths sends them smashing into their neighbors. The gravity of the star tugs on them as does the gravity of all the pebbles and boulders around them, pulling in everchanging directions. Imagine you're watching a marathon where contestants are running shoulder to shoulder. If someone stumbles, the other runners will change their paths and bump into the runners next to them, setting off a chain reaction of runners slamming into each other. Now imagine, instead of runners, rocks and snowballs. These pieces sometimes crash into each other and can stick together—unlike the runners. Some fragments are also catapulted out of the disk into the darkness of space or toward the fledging star. These disks survive for only about ten million to one hundred million years, a blink of an eye in cosmic terms. This growth spurt from tiny grains to planets takes only a few million years and allows specks of dust to grow into majestic planets, like our Earth.

Earth is located a minute cosmic distance of about 93 million miles (~ 150 million km) from the Sun. The inner part of the revolving disk is closer to the hot star so that most ice and gas evaporates, leaving mostly rocks behind. Thus, like Earth, the other worlds that formed close to the Sun (Mercury, Venus, and Mars) are rocky, because that was the material available. But farther away, in the outer reaches of the disk—beyond what astronomers call the *ice line*—gas, ice, and rocks smashed together, creating much larger gas and ice giants (Jupiter, Saturn, Uranus, and Neptune). These planets are utterly different than the closer-to-the-Sun rocky worlds.

Imagine a big cosmic bathtub—if you threw in Earth, it would sink like a stone because it is basically a rock. If you threw in Saturn—the best-known gas planet in our solar system, with its beautiful shiny rings—it would float, because Saturn's mean density is lower than that of water. (You can get the mean density of any object by dividing its mass by its volume.) Saturn is about the same density as cotton candy.

Just imagine a new section in Roald Dahl's 1964 book *Charlie and the Chocolate Factory* in which characters could walk on a cotton-candy Saturn in Mr. Wonka's chocolate factory. But Saturn is not made of cotton candy. It is made mostly of hydrogen gas and it is nowhere near as tasty. So mean density alone does not tell the whole story because even hot gas changes under enough pressure.

The farther you dive into a giant planet's atmosphere, the higher the pressure from the overlying gas on the regions below, just as the deeper you dive into the ocean, the higher the pressure. If you took a step onto Jupiter, you would sink into the swirling gas and keep sinking until you were crushed by the intense pressure. The gas gets denser and denser under the increasing pressure until it becomes liquid. So, Jupiter holds the record for the largest ocean in the Solar System. And in the center, Jupiter's core is thought to be solid, at an astounding 45,000 °F (~ 25,000 °C). The pressure there is estimated to be an astonishing 44 million times that on Earth's surface. It would feel like having about 150,000 cars stacked on top of you. (No one has been to the core of Jupiter to confirm this.)

However mysterious the majestic gas planets in our solar system are, Earth, the largest rock circling the Sun, holds the biggest mystery: it is the only world we know of that harbors billions of life-forms.

How to Make an Earth

It all started about four and a half billion years ago, when our planet resembled a nightmare out of a science-fiction book. Earth took shape in a collision of space rocks. The hot young Earth was covered with roiling magma oceans relentlessly bombarded by rocks from outer space. The vast barrage transferred such enormous energy to the fledgling Earth that its surface melted and remodeled again and again, forming a rough black landscape with fresh lava flows ejected from incandescent orange cracks, all of it blanketed by a thick layer of steam. The young Earth was wrapped in a dense atmosphere of water vapor, nitrogen gas, and carbon dioxide. Its air would have been toxic to any unlucky human time traveler stranded there. Even the sky would have been breathtakingly alien. The familiar constellations were missing. All stars move, and the constellations that guide us home today have changed over millions of years. If you stood on that nascent Earth, there would be no Moon for you to admire yet, just a dark and eerily unfamiliar sky.

But then, shortly after, a violent collision changed the alien sky into a more familiar one. A Mars-sized planet—named Theia by astronomers—probably traveled along almost the same orbit as the Earth, and because two different planets can't occupy the same distance from a star, eventually they collided. The powerful impact melted most of the smashed-together Earth-Theia body, and part of that molten rock was catapulted away from Earth. But most of it did not escape Earth's gravitational embrace. The cooling pieces were captured in a ring around our planet and kept colliding with each other until they formed a huge rock. That rock became Earth's companion: the Moon. None of the other rocky planets in our solar system has such a big moon. Mercury and Venus don't have moons at all; Mars has two, but they're extremely small and are likely pieces of the initial solar system—asteroids—that came too close to Mars and were caught in its gravity. Alternatively they might also be the result of a collision early in Mars' history, more observations will tell.

Nowadays, planetary collisions are unlikely. Just think about how much empty space there is between the planets and how difficult it is to hit a small planet in the vast expanse of the solar system. Planetary missions require careful route planning, and they undertake many maneuvers to avoid missing their targets. A rock sailing through space has a much greater chance of missing a planet than hitting one. But the young solar system was very different back then, not all objects had found their final places yet. Many small pieces bombarded the planets in the first few hundred million years; astronomers call this the *Late Heavy Bombardment*. Some collided with each other.

Some collisions were so forceful that the objects splintered into tiny pieces rather than sticking together. These ancient collisions also left us presents: small pieces of material from the time of the birth of the solar system that may burn up in our atmosphere in a beautiful spectacle of light—or reach the ground if they are big enough.

How do we know that Earth is four and a half billion years old? The evidence is contained in a few unassuming rocks that you can see in museums worldwide as rocks that reached Earth's surface from outer space, small shards of asteroids and comets: *meteorites*. You can figure out how old meteorites are because they contain radioactive material. As mentioned briefly before, radioactive atoms spontaneously change into daughter atoms at a known rate. The time it takes for half the atoms to decay is specific for each element, its half-life, So if you measure the number of atoms that have yet to decay compared to the number of atoms that have already, you have an exceptionally accurate clock. This predictable decay rate of radioactive atoms such as isotopes of uranium, potassium, rubidium, and carbon can measure time accurately over billions of years. It solves the mystery of how old meteorites are: they're between 4.58 and 4.53 billion years old. That also tells us the age of Earth because it formed from these rocks circling the young Sun.

The *meteor* showers we see every year are displays of thousands of smaller space rocks entering our atmosphere when the Earth plows through them on its path around the Sun. As these rocks travel through our air, friction with the particles in their path creates so much heat that all of them luckily evaporate long before they reach the surface; we commonly call these *shooting stars* because of the beautiful streaks they paint on the night sky, but they are really ancient rocks burning up before our eyes.

I love to watch these ancient messengers. The darker your surroundings, the more of them you get to see. In August 2019, I was waiting for those small streaks of light to cross the velvet sky in the dark-sky region, close to Alqueva Lake in Alentejo in the southeast of Portugal. The day had been so hot that the wind felt like a hair dryer blasting in my face. Darkness had settled entirely over the area; the fruit trees no longer cast shadows on the walls, but the smell of wildflowers still permeated the air. All the lights were turned off. The only sounds were bats flying through the night. The day's heat had baked the stones under my back, which made for a warm, although hard, bed. The steady, beautiful view of the stars illuminating the dark was disturbed only by the shooting stars of the incredible annual Perseid meteor shower. Some streaks were short and thin. Some were nearly too dim to catch. Some left bright luminous traces. Lying there engulfed in the beauty of the night sky, I envisioned the rocks entering our atmosphere and crashing against the atoms and molecules that make up our air, slowing their ancient journey and generating so much friction that the billion-year-old record-keepers of our solar system's history fall apart and evaporate, leaving only a beautiful streak of light behind as testimony to their journey. But some of these ancient messengers generate more than just a beautiful display. Some hit Earth and they can have a devastating impact. Meteorites, mentioned earlier, can leave scars.

We don't see most of these scars on Earth's surface anymore because on our planet, craters don't last long. Over millions of years, weathering by wind and water reliably erases the history of impacts. Seeing impact craters on a rocky planet means that the bombardment happened recently or the surface has not changed since then. Mercury and the Moon are two examples of surfaces locked in time. You can still see evidence of devastating hits from the Late Heavy Bombardment Period in the young solar system.

Two famous and more recent hits on Earth are the Allende and the Canyon Diablo meteorites. In 1969, the

Allende meteorite entered Earth's atmosphere as a spectacular fireball, exploded, and scattered material around the Pueblito de Allende in northern Mexico. Although no one was around to record the Canyon Diablo meteorite when it hit fifty thousand years ago, that was likely a spectacle as well—it formed Meteor Crater in Arizona. At this point you are likely puzzled because meteors burn up in Earth's atmosphere, they don't strike the ground and leave huge craters. Its misleading name is based on historical convention: the United States Board on Geographic Names recognizes names of natural features derived from the nearest *post office*, so this feature acquired the name of "Meteor Crater" in 1906 from the closest one, located at a railroad flag stop five miles north of the crater, named Meteor.

One sunny day, I hiked to the sweltering bottom of Meteor Crater. The people who had sensibly decided to walk along the rim were only small dots when I looked up at them from 560 feet (~ 170 meters) below. It is astonishing to reach the bottom and confront the 3,900-foot (~ 1,200 meters) expanse of the crater. It makes it easy to imagine the destruction one meteorite can cause. Most of the meteorite's material evaporated in the intense heat created when it crashed into Earth, but the powerful shock wave that resulted wreaked havoc for hundreds of miles around, flattening and incinerating large areas. These craters on Earth seem like messengers from ancient times, but these two hits were by no means the only ones.

If you ever wonder why it's important to explore space—other than to satisfy your curiosity—one critical reason is that Earth is not an isolated globe under a protective cover. It is part of our solar system and embedded in the cosmos. By exploring the space around us, we learn about that place and our own planet and how to counter dangers in our environment. About sixty-six million years ago, an asteroid impact ended the reign of the dinosaurs. But smaller space rocks, like the one responsible for Meteor Crater in Arizona, could cause devastation if they hit a city today. We are lucky that most of Earth's surface is covered by oceans in which no one lives, and only small parts of the continents are highly populated. But if you want to avoid a spectacular end to your civilization, you need a space program. We have to see dangers coming in order to react. To do this, we need to map the sky with telescopes and find space rocks before they collide with Earth. That effort has started already. Yes, a devastating hit is somewhat unlikely, but I'm sure the dinosaurs wished they'd come up with a way to deflect the asteroid that wiped them out.

Humans have recently developed the first tool to escape extinction. Launched in 2021, NASA's DART (Double Asteroid Redirection Test) mission purposely collided with the small asteroid Dimorphos, which orbits a larger asteroid called Didymos, on September 26, 2022. The impact caused Dimorphos to change its orbit around its mother rock, demonstrating that humanity can nudge away an asteroid that's heading for us. The DART mission is the first time in the history of Earth that humans tested a technology that could save us from extinction. I imagine the ghosts of billions of dinosaurs are cheering us on—go, humanity!

Heaven or Hell: What the Planets in Our Solar System Teach Us

Our solar system is full of diverse, fascinating worlds, from the smoldering acidic surfaces blanketed by thick clouds on Venus to the permanent hurricanes on colossal Jupiter to the frozen landscapes on icy moons. They give us a glimpse of just how different other worlds can be.

To see all eight planets and the hundreds of moons in our solar system, we need to zoom out from Earth. The Sun, our star, is located in the middle of the eight planets circling it. Our solar system also contains what looks from a distance like two belts: the asteroid belt, which is located between Mars and Jupiter, and the Kuiper Belt, which lies beyond Neptune's orbit. Both doughnut-shaped regions are filled with millions of primordial rock-and-ice pieces, from small meteorites, to large asteroids, along with several bigger pieces we call *dwarf planets*. Dwarf planets, like Pluto and Eris and dozens of others, move around the Sun with millions of smaller rock-and-ice pieces scattered in the belts. Several spacecraft are mapping these ancient, primordial objects, collecting photos and even some material to give us further glimpses into the spectacle that was the birth of our solar system.

Let's start with an inventory: Mercury, the innermost rocky planet, resembles a heavy, scorched core. Our smallest planet probably lost most of its outer layers in a scarring collision early in its existence. Each day, ever so slowly, daylight and boiling heat creep over its surface reaching up to 800 °F ($\sim$ 430 °C), following a devastatingly cold night of −290 °F (−180 °C). Mercury does not have an atmosphere that could mitigate the temperature extremes.

The next rocky world, Venus, is a near twin to Earth but completely shrouded in sulfuric acid clouds. Venus is the hottest planet in our solar system; it's even hotter than Mercury, which lies closer to the Sun. Venus has a similar mass and size as Earth, but whereas Earth is a paradise teeming with life, Venus's roasting, desolate landscape is inhospitable to living things, with surging surface temperatures surging to 900 °F (∼ 480 °C), hot enough to melt lead.

After Earth, the next world is a red rock, Mars. About half the diameter of Earth, Mars features volcanic landscapes in hues of red, orange, and brown. With a thin atmosphere that does not trap much heat, temperatures range from about −230 °F (−150 °C) to 70 °F (∼ 20 °C). Its surface shows evidence of ancient water: braided channels, fan-shaped deltas, clays, and minerals that form in contact with water. This water was either lost or trapped in ice as Mars became older and colder.

Depending on where Mars and Earth are in their trip around the Sun, the signal from Earth to Mars can take between four (when they are closest to each other), and twenty minutes (when they are on opposite sides of the Sun). So telling a Mars rover not to drive off a cliff could come nearly forty minutes too late: up to twenty minutes until mission control gets the images and sees the cliff and up to another twenty minutes to tell the rover to stop. If you ever wondered why Mars rovers drive so slowly—their maximum speed has been about 0.1 miles (∼ 0.15 km) per hour—it is because rovers need to assess their environment independently, spotting imminent danger without Earth's help.

Between Mars and Jupiter, you find a wide ring of icy rocks that seem to have forgotten to form a planet: the asteroid belt. Farther out lies the realm of the giant gas and ice planets, which contain most of the material left over from the Sun's creation and dwarf the nearer rocky worlds. Jupiter, the majestic planet governed by enormous storms; Saturn, with its brilliant rings; stormy Uranus and Neptune; these worlds were all forged of gas, ice, and rocks. They circle our star much more slowly than the inner planets because the Sun's gravity loosens its hold the farther away a planet is. Planets farther out can circle leisurely and still balance the Sun's weaker gravitational embrace. Beyond Neptune, we find the second belt in our solar system, the Kuiper Belt, which contains millions of small rock-and-ice pieces and is home to dwarf planets like Pluto and Eris.

Even farther out, we encounter a cocoon around the planets and belts—the *Oort Cloud*. A vast region that lies between two thousand and one hundred thousand times farther out from the Sun than our Earth. It is a spherical layer of billions of small icy and rocky objects. Even at that distance, the Oort Cloud is caught in the Sun's gravity and therefore is part of our solar system. Light from the Oort Cloud takes between eleven days and one and a half years to reach us.

Let's dip into the realm of science-fiction plots and imagine a cosmic supervillain that needs water: stealing it from the outer parts of the Oort Cloud rather than from Earth would be a smart choice. We wouldn't even know about it until a year later. Quite a good plan if you want an excellent getaway window. And this strategy would avoid any conflict because we would not miss Oort Cloud water.

Remember the Golden Records on route to the stars? The signal from Voyager 1 takes more than 22 hours and Voyager 2 more than 18 hours to reach Earth from their distant vantage points now. But even though the Voyager missions travel close to a speed of one million miles (~ 1.6 million km) per day, it will take more than 25,000 years before the spacecraft exit the far side of the Oort Cloud. Both craft have cleared the *heliosphere*, the protective bubble created by the Sun that shields the solar system planets from interstellar radiation. The probes have left the protection, but not the gravitational embrace of our Sun. Their scientific missions end when their plutonium-238 power generators fail, which is expected around the year 2030, leaving the two Voyager spacecraft and our message from Earth to drift among the stars. While no human-made object has left the solar system yet, our messengers are en route to undertake that spectacular adventure and become our first interstellar explorers.

But back to how to build a habitable world.

Second: Add Energy

Once we have a planet, another ingredient we need for life is energy. There are many energy sources, but on Earth, the largest input comes from the Sun. As our closest star, the Sun has shaped our beautiful planet tremendously.

Understanding our star is key to understanding its interaction with our world.

When you watch the night sky, you can tell that the Earth is moving. Most stars seem to rise and set during the night, because the Earth rotates around its axis every twenty-four hours. It also makes the Sun rise every morning.

The Sun is so close to us that it outshines all other stars in the sky. If you look at a flashlight from nearby, then from far away, you can see that the light appears brighter the closer you are. No matter how many other dim lights shine in the background, you won't be able to spot them over the glare of the bright flashlight in front of you. We can see the other stars only at night, when we are not on the side of Earth that is bathed in sunlight. The stars are always there, even during the day. You just can't see them because their light is outmatched by the brightest star in our sky—our Sun. Unless they explode, but more on the death throes of stars later.

The stars you can see at different seasons in the night sky change because the Earth moves on its path around the Sun. Stars that together seem to form distinct shapes helped explorers find their way on a dark night. Well-known constellations, like Orion, are patterns our brains create from similarly bright but otherwise mostly unrelated stars. From Upstate New York the constellation Orion, a group of stars that the Greeks thought looked like a hunter, is visible high in the sky every late summer to autumn, but not in spring. We see the same stars in a specific season every year because the Earth is at the same place on its path around the Sun. (For the few lights that move unusually, the Greeks used the word *planetes*, meaning "wanderer.") Stars, including those that make up the beautiful constellations you can find on any star map, stay pretty much where they are over the course of a year. But Earth's location on its path around the Sun changes, and that determines which stars we can see in the night sky.

You are probably more familiar with this concept than you realize: imagine a straight line from Earth through the Sun far out into space. This imaginary line points to different stars while the Earth completes its path around the Sun along the *ecliptic*, the imaginary plane of the Earth's orbit around the Sun. About 3,000 years ago Babylonian astronomers designated 13 constellations that this imaginary line points to in its yearlong journey: Capricorn, Aquarius, Pisces, Aries, Taurus, Gemini, Cancer, Leo, Virgo, Libra, Scorpius, Sagittarius, and Ophiuchus. You are likely familiar with 12 of these constellations, the ones the Babylonians picked to fit their established ancient calendar of 12 lunar months. They left out Ophiuchus (end of November), the serpent bearer or snake charmer, probably for convenience. That is why we have the Zodiac of today.

Most ancient people assumed all objects in the sky moved around the Earth, so it seemed obvious that the cosmic forces could influence human beings. Only with the invention of the telescope in the seventeenth century came the realizations that the Earth orbits the Sun and also that stars in one constellation are generally only linked by line of sight but not distance—so do not really belong together—meaning this idea started to fade.

Our Sun is one of about two hundred billion stars in

our Milky Way. And only one of billions of galaxies are out there, some with more and some with fewer stars. The estimates are around one trillion stars for the biggest galaxies and one million stars for the smallest. Imagine the enormous pile of sand you'd accumulate if you put all the sand of all of Earth's beaches together . . . and there are still more stars out there in the cosmos than the number of grains of sand in that pile.

New stars are continuously forming in clouds of gas across the universe, like the Eta Carinae Nebula, which the JWST has observed in exquisite detail. In these stellar nurseries, thousands of stars are formed at the same time. A star is "born" when it starts fusing the lightest element, hydrogen, to helium in its core. The Sun began doing this about four and a half billion years ago and will keep doing it for another six billion years. How do we know? By observing the sky, astronomers pieced together how stars form, grow, and die, expanding our horizon billions of years beyond humankind's existence into the past and the future. Science is and always has been an adventure of discovery.

Light needs about eight minutes to travel from the Sun to Earth. If our Sun were to explode, the sunlight would not vanish until eight minutes after the grand explosion. Being closer to the Sun, Venus would receive sunlight for only about six minutes; Mars, about one and a half times as far from our Sun as Earth, would enjoy about twelve minutes of sunshine, four minutes more than our planet. Those final eight minutes of sunshine on Earth would probably go unnoticed. No one would know that the Sun would never shine again. We would realize the Sun was gone only when the news of the explosion reached our planet—in the form of the absent sunlight—eight minutes later. My students really appreciate sunlight after we cover this topic in class. But don't worry—nothing indicates that the Sun will explode in the next few billion years.

If you looked at our neighboring star Proxima Centauri through a telescope tonight, the light you would see was sent out when a child who is four years old today was born. So if the mass of glowing hot gas that is Proxima Centauri exploded, you would not know about it for four years. But Proxima Centauri has a much longer lease on life than even our Sun. It is a red dwarf star, much smaller than our Sun, and very long-lived. It is also part of a triple-star system— three stars circling one another in a dance orchestrated by gravity. We'll talk more about small stars and our stellar neighbor a bit later. Polaris, the North Star, is about three hundred light-years away from Earth, so the light that guides you tonight was sent out roughly three hundred years ago.

Because light needs time to travel through the cosmos, you can find a link to your own past in the sky. There is a star in the night sky whose light was sent out when you were born and is just arriving now. Making this cosmic connection can be a special present for yourself, for a friend, or for a loved one. Go to your browser, type in *star* and the age of the person of interest, then add *light-years away*, and you can find the stars in the night sky that are linked to the time of someone's birth or any other special time in their lives. This works for anything that happened more than four years ago, because four light-years is the distance to our closest stellar neighbor. And your birthday star changes for every future birthday because your age will increase, linking you to stars ever farther and farther away. A different star every year connects you personally to the cosmos.

When you look up at the night sky, you are watching events that have already happened. That means you can still see the light of a star that stopped shining long ago. So is anything we see in our sky definitely still there? Well, as far as I know, the Sun is still there—ask me again in eight minutes.

In their cores, stars fuse elements from hydrogen to helium to carbon and oxygen, silicon, and heavier elements all the way up to iron. To fuse heavier elements, the temperature and pressure in a star's core needs to increase, pushing more energy to the outer layers, which then expand, making the star swell into a huge red or blue giant. Every second, our Sun converts hundreds of millions of tons of hydrogen to helium, generating energy. Due to its mass and the resulting temperature and pressure in its core, our Sun will become a red giant and end its life with a carbon-oxygen heart. Its core won't get hot enough to fuse those into any heavier elements. Once our Sun has exhausted nuclear fusion in the core, it will expel its outer layers—up to half its mass—into the cosmos, where this will become material for new stars and planets, continuing the circle of (star) life.

The Sun will leave behind its stellar corpse: a small, extremely hot, still glowing, dead stellar husk, a white dwarf. A white dwarf is about the size of our Earth, but it's completely different in every other way—it is much, much denser and hotter. Because white dwarfs start out so hot, astronomers can watch these stellar corpses for several billion years after they are revealed. But eventually, they will grow cooler and cooler and finally so cold that they will just fade from view. More about white dwarfs later.

For about ten thousand years, the shell of gas and dust ejected from a dying star shines brightly, lit by the exposed core of the star. These beautiful shells are called *planetary nebulae* because they looked round, like planets, to French astronomers Charles Messier in 1774 and Antoine Darquier de Pellepoix in 1779, who commented that the Ring Nebula in the northern constellation of Lyra looked like a fading planet. As the nature of nebulae was unknown then, the name *planetary nebula* stuck. The JWST captured a stunning new image of the beautiful Ring Nebula. It revealed that another star circles the white dwarf at the nebula's core. The pair at the center of the Ring Nebula stir the pot of gas and dust, introducing asymmetrical patterns in the shells, glowing and expanding into the darkness of space about two thousand light-years away from Earth.

Not all fuel can create energy in a star's core: the most massive stars will end up with a hot iron heart. Iron is a dead end for a star's engine. To fuse iron would require *adding* energy. Once the core of a star is iron, it cuts off the energy production that pushed back against the massive gravitational pressure of the overlying material, leading to the collapse of the core.

Imagine the stars tonight scattered across the canvas of our nighttime sky. Among them, occasionally, a massive star undergoes an extraordinary transformation. In a sudden cosmic crescendo, the star collapses inward, before rebounding in a violent, explosive performance that outshines entire galaxies, which can become visible in the sky even during daylight, a *supernova*. Stars more than about eight times as massive as our Sun die in these powerful cosmic explosions of energy and light, atomic nuclei and exotic particles. But the finale of this performance is not the end; rather, it's the birth of something new. The death throes of massive stars seed the cosmos with the elements necessary for new stars and planets, including the six essential elements—carbon, hydrogen, nitrogen, oxygen, phosphorus, and sulfur (called the CHNOPS elements)—required by all life on Earth.

Supernovas are beacons of light that you can see over large cosmic distances and some types always shine with the same brightness. Because of this, they are one of the yard-sticks that let astronomers map the universe. The last time you could see one with the naked eye here on Earth was in February 1987: Supernova 1987A (*A* indicates the first one of that year). The blue supergiant Sanduleak −69°202 exploded 168,000 light-years away from Earth. The super-nova shone so brightly you could see it just by looking up at the night sky. But Supernova 1987A happened about 168,000 years earlier. Its light didn't catch up to our cosmic history book until 1987. Supernova 1987A did not even occur in our own galaxy. Sanduleak −69°202 was part of the Large Magellanic Cloud, a satellite galaxy of our own Milky Way, but its final explosion sent a beacon to neigh-boring galaxies. Supernovas happen only about three times per century in a galaxy like ours. So even before we discuss our star's future in more detail, it is an excellent bet to say the Sun will still be there tomorrow morning.

Currently, the Gaia spacecraft of the European Space

Agency (ESA), launched in 2013, pinpoints the positions and the movements of billions of stars with exquisite precision, offering us a detailed view of the elegant gravitational dance of stars in our galaxy. It chronicles the movement of those stars around an enormous black hole with astonishing density at its center. A black hole generates such an immense gravitational pull that even light can't escape. Think of its immense gravity as a spiderweb—anything that gets too close is trapped. That is why astronomers call it a black hole. If no light can escape from the inside of a black hole, no information is carried out of it to tell us what it is like inside this strangely fascinating singularity. Most galaxies have massive black holes in their centers, but those are bigger than any that current stellar explosions can create. They paint an intriguing picture of an assembly of massive stars in a young universe creating enormous black holes that then merged into even bigger ones that now form the heart of most galaxies.

Our Sun circles the black hole, Sagittarius A*, in the center of our Milky Way. We are about 25,000 light-years away from Sagittarius A*, but like all stars in the Milky Way, our Sun and its planets are caught in its gravitational embrace. Monty Python's "Galaxy Song" provides a fun and pretty accurate summary of our movement in the universe—and asks hopefully, in satiric Monty Python style, whether there could be intelligent life in the cosmos because, they lament, "there's bugger all down here on Earth."

One complete circle around our galactic center takes about 230 million years, marking one galactic year. To put this immense timescale in perspective, the dinosaurs roamed

Earth from about 250 million to about 66 million years ago. So we have something in common with the dinosaurs: we are just starting to travel through the part of the cosmos that the Earth traveled through when the dinosaurs were here. The solar system whizzes with an astonishing speed of more than 500,000 miles (800,000 km) per hour around the galactic center to complete one circle in a mere 230 million Earth years, so if you ever feel stuck, remember: cosmologically speaking, you are not. You are speeding through the cosmos. And you are part of it.

Except for hydrogen, helium, and small amounts of lithium and beryllium, all the elements that make *you* were produced in the inferno found in the cores of stars or during their violent death throes in spectacular supernova explosions. Elements heavier than iron, like silver, and gold, are forged in the immense supernova explosion at the end of a massive star's life. You can hold the leftover fragments forged in a star's death throes in your hand. The calcium in your bones, the iron in your blood, and the oxygen you breathe are all ancient stardust. In the vast expanse of the universe, you are part of the cosmos.

You are made of ancient stardust.

Will Life End in Fire or in Ice?

Just as we seek to understand the conditions under which life might evolve on a faraway planet, we wonder how life there might end. It turns out that it depends on the color of its parent star. For Earth, the world ends in fire. In about six billion years—so we have a long time left—our Sun will run out of hydrogen to fuse in its core. Then the Sun will swell to about two hundred times its size, becoming a red giant, that will engulf Mercury, Venus, and maybe Earth (this is still debated; Earth might just escape engulfment). Even if Earth doesn't become part of the Sun, our Sun's energy will incinerate Earth's surface. But at that point—hopefully—we will already be traveling among the stars.

What about our neighboring stars? The backbone of our understanding of our place in the cosmos comes from decades of observations; astronomers have painstakingly measured the brightness and precise positions of thousands of stars. If you were to do an inventory of all the stars in our neighborhood, let's say within thirty light-years, would these stars be the same as our Sun? Imagine a group of ten stars representing these roughly four hundred stars. There are eight small, cool, red stars in this group, one coolish orange star, and one strange-looking star made of two-thirds yellow sun and one-third hotter white star. Small red stars are by far the most common. But what does a star's color mean?

The heat in the stellar core controls the incandescence of its surface, which in turn determines the star's color. Counterintuitively, *redder* means cooler surfaces; *bluer* means hotter ones. Imagine an iron poker that you heat over a fire: first it glows red, then yellow, then whiteish blue as it heats up. So a yellow star like our Sun is hotter than a red star like Proxima Centauri.

Stars can live for billions of years without significant changes. They shine brightly because deep within a star's core, hydrogen atoms come together under extreme heat and pressure in a dance of nuclear fusion. This merging of atoms creates a delicate equilibrium, balancing the force of gravity that tries to crush the star with the energy released from nuclear fusion. Heavier, more massive stars have higher temperatures and pressures in their cores, making them blow through their nuclear fusion material much faster than smaller red stars do. So the length of time stars survive depends on how massive they are. Yellow stars like our Sun live for about twelve billion years, and luckily, our Sun is not even halfway through its lifetime at the age of about four and a half billion years. Red stars live at least ten times longer; the smallest ones can live hundreds of billions of years. Or so we think. Astronomers have not seen a red star die; the universe is just not old enough yet.

To determine the age of a group of stars that formed together, astronomers count how many massive stars there are in that group compared to how many formed. The fewer, the older the group is. Our Sun can not sure all of its hydrogen, only the part in its center, to generate energy. Even though the Sun is a swirling ball of hot plasma, the hydrogen in the outer layers does not get mixed into the core, so it stays out of reach as additional fuel. Small red stars are different. They are fully convective. That means that their material gets mixed. All the material from the outside gets sucked into the center, where it is hot enough to fuse. But these small red stars can fuse only hydrogen to helium. Their cores do not reach the temperatures and pressures needed to fuse the heavier helium, so they stop producing energy when they run out of hydrogen, fading from view.

Most of our star's travel companions are red stars. For

planets circling those red suns, the story ends completely differently: life ends in ice. Instead of swelling up to the size of Earth's orbit, a red star just stops producing energy once it runs out of hydrogen. Then it gets colder and colder until its light fades away entirely, leaving any Earth-like planet circling it a frozen wasteland. But the differences do not end there. Planets under a red sun have a future that stretches much further than Earth's. Because small red stars have such immense lifetimes, they could provide steady light to their planets and any organisms on them for hundreds of billions of years.

The exploration of new exoplanets on our cosmic horizon, including younger and older rocky planets, will help us uncover the story of the life of rocky worlds. But let's get back to *our* cosmic far future; billions of years from now, our planet will be nearly engulfed by the Sun. Life on planets orbiting any star must eventually leave to survive—or change the evolution of its star. Carl Sagan wrote, "All civilizations become either spacefaring or extinct." It took only about four billion years for life to come to that critical insight here on Earth, circling our medium-mass yellow Sun. If there is life on a planet circling a more massive star, which runs through its nuclear fuel much faster, will there be enough time for any organisms to come to that same realization? In contrast, life on planets under a red sun could enjoy their star billions of years longer than us before it would get ever colder under its dimming light. Imagine civilizations inducing global warming to stretch temperate conditions out under the light of an ever-cooling star.

Which would you prefer: Going out in a hot blaze of

glory or a slowly dimming sun? Here's an alternative: let's become wanderers of this amazing universe. It does not have to end in fire or ice.

Third: Add an Atmosphere

Exploring our neighboring planets, the other rocky worlds in our solar system, yielded the first clues of what shapes the atmosphere of planets. Venus has about the same mass and size as Earth, so it was a prime candidate for a habitable world. It is just a bit closer to the Sun than Earth and gets about twice the energy that we do. That led to the eighteenth-century's fantastical idea of a lush jungle paradise on Venus. But once observations pierced the clouds that shroud its surface, astronomers found an acid-laden, hot hell where no life we know of could survive. The atmosphere on Venus is mostly carbon dioxide (CO_2), a gas that traps the Sun's energy very effectively, heating its surface hundreds of degrees above the boiling point of water, destroying any hope for survival in this hostile environment.

What happened on Venus? Imagine sitting in a car on a hot, sunny summer day with your windows closed. The longer the light hits the car, the hotter it gets inside. Like the closed glass window, CO_2 traps the heat in an atmosphere. While incoming visible light passes freely through our air, heat gets partially trapped by molecules like CO_2 methane, and water, warming a planet like a blanket.

Humans can't see heat (infrared radiation) because our eyes evolved to use the peak energy from the Sun, the visible part of light. We see objects around us because light reflects off them. You and I reflect light, but we glow in the infrared too: it is just not in a wavelength our eyes can see. If we could, we would see warm bodies in the dark. A few animals can see heat; for example, goldfish can because they generally live in murky water. (Mosquitoes can too and use it to unerringly find me in the dark.) Even though our eyes can't see heat naturally, we can use this concept to create technology to help us achieve it. Night-vision goggles use thermal-imaging technology to capture infrared light and translate it to visible light so you can see heat and use it, for example, to find other humans in the dark.

The balance between incoming stellar radiation and outgoing heat on a planet creates the temperature on the surface. Like the different interactions of glass with visible and infrared light allows sunlight into a greenhouse (or a car) but keeps the heat from escaping, gases can trap heat in the atmosphere. These are called *greenhouse gases* after the greenhouse effect they enable. In 1896, the Nobel Prize–winning Swedish chemist Svante Arrhenius discovered how air warms the Earth. And he realized that the additional human-made CO_2 emissions were already influencing Earth's temperature. But there was no alarm yet. Arrhenius even popularized the idea of a strikingly warm, wet atmosphere on Venus. It took decades before Carl Sagan realized that Arrhenius had laid the groundwork to solve the mystery of Venus's scorching conditions.

Sagan had the critical insight that what was true for Earth must also be true for other planets: while each planet has its unique story, the same basic forces shape its climate. Sagan took what we know about CO_2 on Earth and applied it to the thick Venusian atmosphere. His research led him to the realization that the CO_2 on Venus was trapping enough energy to turn our neighboring planet into an infernal world. Several spacecraft visited Venus, measured its atmosphere and surface with cloud-penetrating radar, and confirmed this. The Soviet Union's space agency managed to land on Venus's surface in December 1970. Despite the Venus-proofing design efforts to enable eight of its Venera landers to successfully touch down on the surface, the landers survived at most a little over two hours before succumbing to the intense heat and crushing air pressure.

That Venus is so different from Earth leads to an uncomfortable existential question: What is the tipping point at which a world can no longer combat the greenhouse effect? Was Venus doomed from the start, or did it provide a paradise for a short time before the heat on its surface evaporated any oceans and made it hostile to life (at least life as we know it)? New large international missions are planned to fly to Venus in the early 2030s to learn more about the fate of our neighboring world: NASA's DAVINCI and ESA's EnVision Mission. In addition, an American space company, Rocket Lab Ltd., in collaboration with MIT is planning a small private mission to Venus.

On Earth, CO_2, luckily, accounts for less than 1 percent of our air, warming the surface by about 60 °F ($\sim$ 30 °C). Not all CO_2 is bad. If we had no CO_2, most of Earth's surface would be frozen. But Venus shows the catastrophic effects of too much CO_2: a runaway greenhouse effect, where CO_2 heats a planet way beyond the temperatures for liquid water. It points to a possible future in which our thriving planet becomes a barren wasteland—the real twin of Venus—with no evidence left of a civilization that was starting to explore the cosmos.

When I look up at the sky now, I imagine the trade winds, those giant rivers of air driven by the Sun's heat and the Earth's rotation, transporting air around our world overhead. Hot air in the tropics always rises, and cold air at the poles always sinks. The air between fills those gaps, creating a giant equator-to-pole circulation pattern of rivers of air. Earth's rotation twists and bends those giant streams of air, adding a east-to-west component to the flow, creating the trade winds that have propelled sailors for decades.

While these rivers of air are invisible to us, they shape our world's climate; they're like giant highways for air to travel the globe unhindered. Winds like these will also govern other worlds orbiting distant stars. Imagine some of these worlds showing these rivers of air in swirling colors, creating a breathtaking spectacle in the sky.

Just Right? The Goldilocks Zone

Our neighboring planets don't host life, as far as we can tell. What made Earth special? The answer is likely water. (We will discuss why all life on Earth needs liquid water a bit later.) Surprisingly, less than three percent of all water on Earth is fresh water and most of it is locked in ice in glaciers or snowfields. Imagine the total amount of water on Earth as filling a gallon jug: all the fresh water on Earth would only amount to less than half a cup of ice cubes. And all the freshwater we have access to on the surface would be only a few drops off these frozen cubes.

Salty or fresh, water is critical to life on Earth. Life uses water as its solvent, a powerful medium that dissolves other substances. We can use life's need for liquid water to determine the best cosmic real estate for life as we know it. NASA developed the slogan "Just follow the water" based on this insight. To keep rivers and oceans glistening on the surface, a planet needs to be at the right distance from its star, not too hot (like Venus) and not too cold (like Mars), the habitable zone or the Goldilocks zone as introduced earlier. This is where conditions are "just right" for liquid water to flow. Imagine a bonfire on a cold night. To keep warm, you want to be close to it, but not too close, or you'll get uncomfortably hot. The right distance depends on the size of the bonfire. If the bonfire is tiny, you want to be very close. If the bonfire is enormous, you'd rather be farther away. There is a region around every star where the surface of a rocky planet would get just the right amount of heat—not too much and not too little—for rivers to flow. To be accurate, it should really be called "the liquid-water-on-the-surface zone." But I agree, that's not catchy at all.

The habitable zone in our solar system extends roughly from Venus to a bit farther out than Mars. Earth is right in the middle of it. Venus, being closer, received about 70 percent more heat from the young Sun than modern Earth does now. Young Venus became so hot that if it had any oceans, they evaporated, leaving behind an arid wasteland. Or maybe Venus was always too hot, and liquid water never accumulated in oceans there—the discussion is ongoing. Perhaps, just for a little while, it was the paradise that eighteenth-century poets imagined. A young Mars received about 70 percent less heat from the Sun than modern Earth. With so little energy from the Sun, once Mars's core cooled, water became locked into permafrost and Mars lost its ability to recycle and build up greenhouse gases, leaving the planet cold and dry. Mars is in the right place for life, but you're still out of luck if you're looking for ocean views, proving that location is not everything. Size matters too—or really the inside of the planet matters. A world in the habitable zone is not automatically habitable, and a world outside it is not necessarily uninhabitable. But outside the habitable zone, life will be even harder to discover because a huge ice layer covering oceans where life could be thriving will hinder our exploration by blocking our view. On Earth we can drill through the ice to check if there is life underneath. We won't be able to do that on exoplanets, so we concentrate on planets where liquid water can flow on the surface and gases aren't trapped under huge ice sheets, hidden from our telescopes.

Comparing Earth, Mars, and Venus shows that Mars and Venus are missing one of two key ingredients to recycle gases in the air: a molten core that allows for the movement of tectonic plates (Mars) and water (Venus). On Earth, plate tectonics is a key part of stabilizing the climate. Neither Mars nor Venus shows evidence of plate movement now, so tectonics is not a given for a rocky planet.

On Earth, volcanic eruptions naturally add CO_2 to the atmosphere. The CO_2 is removed from the atmosphere by weathering (the breakdown of rocks by erosion and reactions with air and water) and pushed back into the mantle, then released into the atmosphere again through volcanic eruptions, creating the *carbonate-silicate cycle* on Earth. Imagine the violent eruptions of volcanoes on a young Earth: massive clouds of gases shooting into the air, darkening the sky until rain cleaned it again. Rainwater and CO_2 form carbonic acid, which can dissolve silicate rocks, the crust of our planet. These dissolved chemicals are carried off by rivers to the oceans. Here, they sink to the ocean floor. Once marine organisms evolved they temporarily get a new lease on life and become shells and skeletons. When death strikes those organisms, the shells and skeletons sink to the ocean floor. On the ocean floor, they get dragged under at subduction zones into the mantle region of the Earth, where they melt and free CO_2 to travel back into Earth's atmosphere through volcanoes. This cycle regulates Earth's atmospheric CO_2 concentration and its greenhouse effect on timescales of millions of years. Unfortunately, that is not fast enough to protect us from the effects of human-triggered climate change, but it was fast enough to keep Earth from being frozen for most of its young life.

The younger Sun was only about 70 percent as bright as it is today, so Earth received only seven of every ten photons it gets now. Today, our planet would freeze over with so little energy coming in. Scientists call that the *faint young Sun* problem. But surprisingly, the Earth was not frozen solid when it was younger. Large amounts of greenhouse gases like CO_2 blanketed our young planet, keeping it warm. As the Sun's brightness increased, the incoming energy increased. That changed the chemical makeup of the atmosphere of Earth, with the carbonate-silicate cycle regulating the surface temperature.

But this cycle is not a magic wand saving all rocky planets. Mars shows us that a planet must be massive enough to keep its core molten; Mars wasn't, so its innards solidified and volcanoes grew dormant and stopped spewing out greenhouse gases, resulting in a cooling planet.

Something else went wrong on Venus. Venus is the same size as Earth, so it is massive enough for active geology. But on Venus, the climate cycle is broken—if it ever worked—because the planet lacks a crucial ingredient: liquid water. Once oceans on Venus evaporated billions of years ago, water vapor made its way high up into the atmosphere, initiating the beginning of the end. Harsh high-energy radiation hits the top of all planets' atmospheres. It can break water into its components, hydrogen and oxygen. Hydrogen atoms are the lightest atoms and easily escape a planet's gravitational pull. The water could not re-form, and Venus permanently lost its precious liquid. Less rain meant more CO_2 staying in the atmosphere, which meant higher surface temperatures, which meant more ocean evaporation and more hydrogen getting lost to space, setting in motion a catastrophic water loss that led to the hot, acid wasteland of Venus we see today.

Luckily, Earth's modern atmosphere is very different

from Venus's. The temperature on Earth generally gets colder the higher up you go, but around ten miles (~ 16 km) up, the temperature increases again because high-energy solar ultraviolet (UV) radiation first crashes into oxygen (O_2) in our atmosphere and produces ozone (O_3), which forms a layer high up in the atmosphere that blocks the most harmful UV rays. Due to that energy, the temperature gets hotter. This change in temperature acts like a lid on a cooking pot—it blocks most of the water from reaching the top of the atmosphere, and water falls back to the surface as rain or snow. With water trapped effectively, the carbonate-silicate cycle can stabilize our climate. If surface temperatures rise, more water evaporates, which generates more rain, trapping more atmospheric CO_2 in rocks and causing the planet to cool. If temperatures decrease, less rain and part of the rock surface covered by snow and ice mean less weathering, so atmospheric CO_2 concentration increases, warming our planet again. Unfortunately, this cycle works on million-year timescales and won't get us out of our man-made CO_2 increase. We'll have to fix that ourselves.

As our Sun gets more luminous, that handy thermostat will break as it did on Venus. Scientists estimate that Earth's climate will get hotter and hotter until, in about a billion years, higher temperatures will eliminate the cold trap on Earth and nothing will keep water from reaching the top of the atmosphere, where it will set catastrophic water loss in motion. The arid wasteland of Venus is a glimpse of our possible future, but we can influence how long it will take us to get there and perhaps even figure out how to engineer a way out of this coming disaster.

Does Life Need a Moon?

Earth has something most other rocky planets don't have: a large companion, our Moon. For most of human history, the origin of the Moon was a mystery. It took the space age and Moon rocks collected from NASA's Apollo missions for scientists to figure it out. The Apollo astronauts brought back over a third of a ton of rocks and soil, and the composition of these lunar samples is different from what we find on Earth. They contain less water and more materials that form quickly at high temperatures, supporting the intriguing idea that something smashed into Earth when it was young and part of the molten outer layer that was catapulted away became the Moon. This giant impact also tilted our planet's axis by 23 degrees, gifting us the seasons. Without the violent collision that gave us our companion, we might not have freezing winters in Upstate New York, but we would not have spectacular colorful autumns either.

The Moon formed around fifteen thousand miles from Earth's surface, about twice the distance from L.A. to Sydney. But today, the Moon is about fifteen times farther away, more than two hundred thousand miles. Imagine that young Moon. Being so much closer, it would have appeared massive. But that is not the most significant difference between then and now—ongoing active lunar volcanism would have made the Moon appear black, the dark, cooled-lava crust interrupted only by cracks showing the glowing magma underneath, providing an eerie view. Large and dark, it would have dominated the sky.

The young Moon was so close that it created monster

tides on Earth. But not of water. On a scorching young Earth, where rocks were molten, a turbulent magma ocean flexed and bent at the Moon's pull. The resulting large tides of magma swept around a young Earth. In this dance of mutual attraction, the Moon pulled on Earth's massive tidal bulge, slowing Earth's rotation. But the attraction was not one-sided; Earth's gravitation also pulled on the young Moon, creating tides of magma that swept over the largely molten lunar surface. Because the Moon's orbit around Earth took longer than the Earth's rotation, the tidal bulge, with its extra mass, constantly pulled the Moon ahead in its orbit, increasing its distance from Earth. As our planet's days started to get longer, the Moon became more and more distant. The changes slowed once the young Earth and the young Moon cooled into rigid bodies that kept their shapes much better than molten ones. Our sky changed because of the beautiful gravitational dance Earth and the Moon started about four and a half billion years ago.

Any spinning systems in space will keep spinning because there is no friction to slow them. But even cosmic dances obey the rules of physics. The total spin of the Earth-Moon system has not changed much over time, but the give-and-take between the two bodies has; they've gone from a close embrace to spinning far apart. This ancient cosmic dance gave us a gift: it doubled the hours in Earth's day. Even now, ocean tides generated by both the Moon's and the Sun's gravity are increasing the time between sunrise and sunset ever so slightly. Each day stretches out just a little longer than the one before—about two milliseconds each century. That means that in about two hundred million years, I will finally get that extra hour in the day I've always hoped for.

This story of the Moon's origin is based on the latest models of how it formed and how it moves now. Scientists have found additional evidence indicating that the days are gradually increasing by looking at systems with cycles that leave marks on our world. For example, modern corals show about 365 lines for every year of growth, one line per day. But fossil corals from about 400 million years ago show more than 400 lines per year, recording more than 400 days per year, revealing that days were shorter then.

There is something timeless about losing myself walking in the night on quiet streets, exploring the town or city I happen to be in. And wherever I go, the Moon and the stars are my constant companions. It feels a bit like coming to a new place and finding old friends there. When I watch our Moon at night, I imagine it looming much larger, a black silhouette marked by orange cracks of glowing magma, the start of the story of a rapidly spinning young Earth— astonishingly strange and eerily unfamiliar.

It is easy to find my way under the full Moon, but on some nights, moonlight is scarce. The amount of moonlight you get changes every night because the Moon is a mirror. It is not a great mirror—you can't see your image in it— but the Moon does not shine itself; it only reflects the Sun's light. Moonlight is really repurposed sunlight reflected back to us. Sometimes we see all the light that hits the Moon (full Moon), but most times we don't. That is what creates the phases of the Moon. This gravitational dance also caught our companion in tidally locked synchronous rotation with

Earth, always showing the same face to its planet. Since we only see one side of the Moon, an intriguing question surfaced: whether there is a "dark side" of our companion and what it could look like. The tricky part of observations is realizing that what you see depends on where you are.

Imagine the Moon orbiting Earth—and now add the Sun to that image. The Moon is always half illuminated. It has a day side and a night side, just like Earth, but from Earth, we rarely see the whole lit part. There is no permanent"dark side" of our Moon. During a new Moon, the lunar far side we don't see is fully illuminated. When we see a full Moon, the far side is dark. Like on Earth, the dark side of the Moon is the night side; the lunar night just lasts much longer than ours, about two weeks. As on Earth, when the Sun rises on the Moon, it starts the next long lunar day. An entire lunar day-night cycle—the time between sunrises on the same spot on the lunar surface—is about as long as one month on Earth. And intriguingly, the cosmic Earth-Moon dance slowed the rotation of the Moon so that it provides a marvelous view: a bright blue Earth on the dark night sky. But whether you see it depends on your location because Earth does not rise or set on the spectacular lunar sky. The gravitational dance hid Earth from the far side of the Moon, but there you could watch thousands of stars sparkling on a velvet black firmament.

Walking in the moonlight, I try to envision our Moon slowly spiraling away from us. Currently it recedes from sight by about 1.5 inches (3.8 cm) per year. (That is about the same rate at which your fingernails grow.)

Our planet and its organisms evolved under the effect of the Moon, which generates tides and stabilizes Earth's seasons. While the height of the tides would change on a planet depending on whether a moon existed, life in a deep ocean would not be affected. And surface life would evolve for these different conditions. Having a moon or not should not affect habitability on other worlds—but I wonder how many moons a planet could harbor, creating a beautiful spectacle in the sky.

If you were to live on a moon instead of a planet your night sky would show the planet in an ever-changing light as well. Standing on the Moon, you would be able to see Earth phases similar to the lunar phases we observe from our vantage point. And if you take a walk over the lunar surface, you can even see Earth rise above the horizon. *Earthrise* is also the name of a photograph of our Pale Blue Dot taken from lunar orbit on December 24, 1968, by Apollo 8 astronaut William Anders. Much later, he reflected, "We set out to explore the Moon and instead discovered the Earth."

Strangely Familiar: Welcome to an Alien World

In the *Earthrise* photo, our planet looks like a blue marble. But when measuring the Earth, you find that it is actually not a sphere. Because it rotates, it is somewhat squished—the poles are flattened, and it has a slight bulge in the middle. And because the continents are not evenly distributed, Earth's shape is wider below the equator than it is above, a bit like a pear, lovably chubby.

Our planet is still a fascinating puzzle, and only recently

have some of the pieces begun clicking into place. If you could cut the planet open, its interior would look like a slightly misshapen hard-boiled egg. That I think of Earth as an egg now is all Andy Knoll's fault, because that is how he describes our planet in his engaging 2021 account *A Brief History of Earth*. Earth's center, the yolk in this analogy, consists of a solid inner core surrounded by a molten outer core; together, they make up about one-third of Earth's mass. Hotter, less dense material near the base rises, and cooler, denser material sinks to the bottom, generating the electric dynamo that powers Earth's magnetic field.

Unfortunately, no one has managed to journey to the center of the Earth yet, so scientists have used a combination of measurements and laboratory experiments to extrapolate what the core is made of. But although we can't go there, there are other ways to explore Earth's extremely hot core. Scientists study waves of energy created by earthquakes to learn about the interior of our planet. The waves are transmitted, reflected, or absorbed at the core, revealing how large it is, how dense, and what chemicals it is made of. Our planet's core seems mostly to be made of iron. That makes sense because when Earth formed, intensely heated by collisions and the decay of radioactive material, iron would have sunk to the planet's center.

Earth's mantle, the white of the hard-boiled egg, surrounds the core and makes up about two-thirds of Earth's mass. It is solid, but on long timescales, it circulates. Occasionally, the material of the mantle gets transported to the surface. Diamonds are formed a hundred miles or more under the surface, and they usually contain tiny inclusions of mantle material that scientists can study in the lab. If you own a diamond, you have a clue to the history of the Earth.

The part of Earth that we can sample is the crust, the thin eggshell in our analogy. It is only about 1 percent of Earth's mass. It is what's left of an ancient ocean of molten material that spread across our young planet after it formed. The black magma oceans cooled and created the first crust. Tiny mineral grains, called zircons, hold the key to these early times. When zircons form, they include a bit of uranium in their structure, but no lead. Uranium can be radioactive, decaying with a known half-life of billions of years to lead. So, any lead in zircons is a product of this radioactive decay, providing an exquisite clock. The oldest zircon is 4.38 billion years old, nearly as old as Earth itself. Surprisingly, for a world that got such a hot start, liquid water apparently existed more than four billion years ago—at least it did where the zircons formed—manifesting as oxygen signatures in inclusions in the zircons.

Liquid water covers about 70 percent of Earth's surface today. Weirdly shaped continents rise from those oceans—continents that look as if they might fit together like a giant puzzle. In 1912, Alfred Wegener, a German geophysicist, proposed the idea that continents move, an idea that was widely ridiculed at that time despite the continents looking like pieces of a jigsaw puzzle when laid out. For instance, notice how the east coast of South America fits snugly into the west coast of Africa. Wegener did not live to see his theory accepted. He died in 1930 while on an expedition to Greenland. Even when the space age began with the 1957 launch of Sputnik, the world's first artificial satellite, Earth scientists still thought the continents were immobile. Only in 1960, when technology adapted from wartime made it possible to study the Earth more closely, did the idea resurface. Sonar measurements—which use sound to measure distance and direction—discovered mountain ridges under the oceans, and scientists found that, mysteriously, the seafloor close to the mountain ridges was younger than the seafloor farther away. That meant that new ocean crust was produced at the ridges. But if new ocean crust was created in one place, the old crust had to be destroyed somewhere else. Seismometers showed that earthquakes do not happen equally all over the world but tend to occur in specific places. Earthquakes gave the answer to where and how the crust gets recycled: it gets dragged back into the mantle at subduction zones. These zones are the graveyards of Earth's crust, where the material returns to the mantle. When that happens, and the plates that cover Earth's surface grind against each other, it can produce earthquakes around the boundaries of tectonic plates.

Ever since Earth cooled enough for plates to form on the molten interior, they seem to have been moving, tearing continents apart and pushing them together and creating majestic mountain ranges, most of which are lost in time. If you ever find yourself on top of Mount Everest, look for fossil shells, because the summit is made of marine limestone, material that was pushed five miles or more above the sea. Now fossil shells rest high above the tree line as testimony to Earth's transformation. If we run the movement of the plates backward about three hundred million years, we find the continents united in a supercontinent called Pangea.

No Alps, Rocky Mountains, or Himalayas had formed yet; there were no recognizable landscapes anywhere. The rock record tells the tale of at least five supercontinents created and destroyed as the moving landmasses crashed together and tore apart. Continents separated and collided due to the movement of the plates they rode on, changing the face of our world over and over again—and will continue to do so in the future.

Earth is covered by moving plates that are linked to processes deep within the planet. Ocean ridges form where hot mantle material rises to the surface. The sinking crust pulls the ocean floor apart, and new material passively forms at the ridges. Even the solid ground we stand on changes—new ground gets made and old ground gets destroyed every day. The ground underneath my feet keeps moving just a little, steadily year in, year out. The distance between New York and London increases by about one inch every year. When I stand on the shore in New York and watch the ocean, I imagine how the new seafloor is slowly pushing North America and Europe apart. I won't notice the added inch on my flight back to see family and friends, but the knowledge that the world is steadily moving grounds me on crazy days.

When I look out of my plane window, I see a planet shaped by a fascinating mosaic of interacting, interlocking plates. Continents peek out of the oceans on most of them. But why are they above the oceans? The ground you stand on, the continental crust, is made of a different material than the ocean floor is. Granite forms deep in the Earth's mantle and rises up slowly, like warm air in a cold room, allowing it to separate from the denser ocean floor. Granite is lighter than the ocean floor, which is primarily made of denser volcanic basalt. So the continents "float" above the oceanic crust. And water fills in the lower parts, creating oceans with continents poking above the waves. At the subduction zones, the dense slabs of ocean crust plunge deep into the mantle, but low-density granite islands pile up, making larger and larger long-lasting landmasses. Due to this highly effective recycling, the ocean crust is no more than two hundred million years old and so does not record the time before. All that information is buried in the piled-up continents that chronicle Earth's history.

The story of Earth is written in rocks. The older the rocks are, the earlier the chapter but the more they have been altered by heat, pressure, and the dissolving powers of water. Careful analysis lets us read the rock record that holds the secret of our planet's evolution and the interconnected plates that keep rearranging Earth's surface. You can see part of Earth's story in rocks all around you—for example, when you admire the stunning colors of the Grand Canyon at sunset. The canyon stretches through the landscape where the Colorado River carved its path through part of Arizona. These beautiful layered bands of red rock reveal millions of years of geological history. Each layer recorded what it was like at that place when it formed like an incomplete book of the history of our planet, with more and more pages missing the further you go back in time, beautifully interwoven with the emergence of life.

If you could watch our planet through the window of a time machine, you would see the continents wandering across Earth's surface, occasionally colliding, creating mountain ranges and tearing them down again. If you went further back in time, you'd see the oceans gaining more space on the surface and continents losing their prominence. Exactly when the continents first pushed above the water is an open question. But a few billion years ago, in warm shallow ponds on these continents, around hot vents on the ocean floor, or on partially melted ice shelves, the first organisms took shape, changing barren rocks into a world teeming with an incredible diversity of life.

TARDIGRADE

AXOLOTL

CROSSOTA JELLY

BLUE DRAGON: GLAUCUS ATLANTICUS

HORSESHOE CRAB

VAMPIRE SQUID

SCYPHOMEDUSA DEEPSTARIA

PINK SEE-THROUGH FANTASIA

SEA ANGEL

DUMBO OCTOPUS

ALIEN WANNABES?

Chapter 3

What Is Life?

The fact that we live at the bottom of a deep gravity well, on the surface of a gas-covered planet going around a nuclear fireball 90 million miles away, and think this to be normal, is obviously some indication of how skewed our perspective tends to be.

—Douglas Adams, *The Salmon of Doubt: Hitchhiking the Galaxy One Last Time*

Becoming an Exoplaneteer

Corsica is a beautiful French island in the Mediterranean, just off the west coast of Italy. It is well known for its mild climate, gorgeous mountains, and tranquil beaches. In 1998, the first academic conference I ever attended happened to be in Cargèse, in the west of Corsica. I figured that I could stretch my meager budget by taking a last-minute tourist flight and a bus, rather than a taxi, from the airport. Once I made it to the conference—the topic of which was "Planets Outside the Solar System: Theory and Observations"—my room (in an apartment shared with several other students) and meals would be paid for.

"Breakfast" was a croissant and coffee, and "dinner" was often the same, but I would have five glorious days at this international conference. I was in my second year of college in Graz, a small city in the south of Austria, studying both engineering and astronomy, which meant that I attended two universities at the same time: the Graz University of Technology for engineering and the Karl Franzens University of Graz for astronomy. (Education is free in Austria.) Luckily for me, Graz is a small city, and you can bike between the two schools in ten to fifteen minutes, depending on traffic. I loved the combination of fields; it provided a perfect balance for me. In engineering, I was studying the smallest details, from quantum mechanics to fixing electronic circuit boards, and in astronomy, I was learning about the largest structures in the universe, from the shape of galaxies to the framework of the cosmos.

I did not really fit in on the flight to Corsica, sitting in my economy seat mashed between people in hiking gear, but none of the hiking enthusiasts paid much attention to me. Once we landed, I watched the airport empty out and waited patiently for the public bus. The few hours I had to wait could not subdue my spirit. I had so many new things to look forward to. When the bus finally arrived, either fashionably late or on Corsica time, the driver made up for the delay by traveling at breakneck speed along the coast.

To my left was the mountainside; to my right, sharp cliffs with the foaming ocean far below. That hour-long trip was not for the faint of heart. I was desperately hoping that no car would come toward us because the bus driver would never have made it to his side of the road in time. Luck was with us, or maybe everyone on the island knew to keep out of the way of the public bus.

Eventually, I arrived in the sleepy town of Cargèse. The sea breeze carried the smell of salt into the airy conference room where fifty of us met to ponder the first new worlds in the cosmos. The coffee breaks (also known as breakfast, lunch, and dinner) were held outside in the warm sunshine and with a view over the ocean. I felt my world expanding.

Among the people attending the meeting was Didier Queloz. In 1995, he and Michel Mayor had detected the first planet circling another Sun-like star. Queloz had been a thirty-two year-old graduate student when he and Mayor discovered this new world—only eleven years older than I was when I attended the conference. Both Queloz and Mayor are Swiss, and this amazing discovery had been made in that small mountainous country that was a bit like Austria. So maybe there was space in the field for me too?

I vividly remember the conversations on the seashore during the conference breaks, all of us full of the excitement of these new discoveries and with so many questions we had no answers to. Surrounded by professors and scientists from all corners of the world, I found myself part of this new scientific adventure. My comments were heard and considered even though I was only an undergraduate student. To this day, I remember James Kasting, the American scientist who established the limits of the Goldilocks zone, asking me seriously after a talk, "Lisa, what do you think?" The smell of spring was in the air, and here was Jim, a well-known senior scientist in the nascent field of exoplanet search, asking me to share my thoughts on a new idea! This was a novel experience for me. In Austria, my professors seldom asked for the opinions of undergraduate students, but hierarchies were suspended within this international team dreaming of new worlds to explore.

In these few days in Cargèse, my world changed. Several of the speakers mentioned during their talks that they needed help with their research—there was so much to explore and so few people to do it. I wanted to be part of this adventure, part of finding life in the cosmos. And for the first time ever, this was not a faraway dream but a real possibility.

After the conference, I headed back home in a plane stuffed with hiking gear, back to my same life—the same classes, friends, and exams to think about—but my world-view had monumentally shifted. Somewhere, out there, new worlds waited for me to explore them.

How We Got Here

Life on Earth is built on carbon scaffolding, and it uses water as its solvent. The abundance of hydrogen, carbon, and oxygen in the universe means that life on other worlds, if it exists, is likely to be supported by water and carbon.

But let's look at alternatives. What about an element that behaves similarly to carbon, like silicon? Silicon is actually more abundant on Earth than carbon. There are several reasons why carbon seems superior to silicon as the scaffolding for life. One is its ability to form a huge variety of bonds with many atoms, including itself. Carbon can create stable complex molecules, like DNA, but it also can break these molecules without expending too much energy. When carbon binds to oxygen, it creates CO_2, which is mostly in gas form under the conditions found on Earth, and is easily soluble in liquid water and therefore readily available for life to use. When silicon binds to oxygen, it creates silicon dioxide (SiO_2), also known as quartz, the rock that makes up one-fifth of Earth's crust. Silicon dioxide does not exist as a gas except at temperatures above about 4,000 °F (~ 2,000 °C). So while silicon binds to oxygen like carbon does, the resulting product is incredibly hard to break into its component atoms again. Maybe that is why life chose carbon over silicon on Earth.

As far as liquids go, water is special. It occurs on Earth in three phases: solid (as ice), liquid (in oceans, lakes, and rivers), and gas (in the atmosphere). It is a powerful solvent, a medium that surrounds and dissolves more substances than any other known liquid. Water stays liquid over a wide range of temperatures on Earth's surface pressure, from 32 to 212 °F (0 to 100 °C). It can stay liquid at even higher temperatures if it is under higher pressure, and at lower temperatures if it is mixed with many solutes. Saltwater will stay liquid down to −22 °F (−30 °C). Water also offers protection from DNA-damaging UV radiation, a feature that organisms in oceans and rivers take advantage of. Water has another interesting characteristic: when it freezes, the resulting ice is less dense than the liquid—it expands by about 10 percent. In addition, it insulates the water below it and keeps that from freezing, allowing life to survive in the depths of a lake through piercing winters. Temperatures near the ocean floor are almost unchanged year-round.

Scientists are exploring ideas about how other solvents might be used on freezing-cold moons like Titan to create alternative forms of life. Methane could replace water as a liquid under these freezing conditions. One concern is that the low temperature of liquid methane of about −256 °F (−160 °C) might slow down any biochemical reactions too much for any life-form to function effectively and thrive. We have not yet found any life that does not use water and carbon, so for now, carbon and water seem to be required for life—as we know it. That provides us with a starting point as we look for other habitable worlds.

Defining Life

When you look for life, what signs can you search for? What characteristics make something alive? In other words, what is life? It is surprisingly hard to define. For example, you can say that life moves. But so does fire. You can say that life evolves. But a computer virus can evolve. Another criterion is that life reproduces—but does that mean that mules (which are sterile) are not alive? You can see how hard it is to define what we are searching for.

We don't have a good definition for life yet, not one that scientists can agree on. In his engaging book *What Is Life?* British Nobel Prize winner in Physiology or Medicine Paul

Nurse provided insight into the debate and developed three guiding principles to define life: (1) life has the ability to evolve through natural selection; (2) life-forms are bounded, physical entities; and (3) life-forms are chemical, physical, and informational machines. Nurse borrowed the title from an earlier, fascinating book by the Austrian Nobel Prize–winning physicist Erwin Schrödinger from 1944 titled *What is Life?* In it, Schrödinger described the physical aspects a living cell must have, which served as inspiration for American biologist James Watson and British physicist Francis Crick, who discovered the structure of DNA later in 1953.

NASA uses a similar definition in its search for life in the cosmos: "Life is a self-sustaining chemical system capable of Darwinian evolution." But a lively discussion continues as to how to best define what life is and how to find it elsewhere.

Cells are the smallest structural unit that can function independently. Think of a cell as a tiny chemical reactor surrounded by a membrane and containing a library of genetic information. Some of the simplest organisms alive today, tiny archaea, pack everything they need for growth, reproduction, and evolution into a single cell. Simple organic compounds can form naturally in conditions that might have been found on early Earth. In 1952, two American scientists at the University of Chicago, Stanley Miller and Harold Urey, demonstrated that the energy from lightning could have created organic molecules on a young Earth. To prove this, they filled a glass container with carbon dioxide, methane, and water vapor—the proposed components of Earth's ancient atmosphere—and ran a spark through the gas mixture, mimicking lightning. This experiment resulted in brown organic material on the container's inner walls, and the results have been replicated in many laboratories all over the world, including at Cornell by Carl Sagan.

The energy needed for prebiotic chemical reactions would have been readily available on a young Earth—high-energy ultraviolet radiation battered the surface, heat from volcanoes drenched the surroundings, and lightning cut through the early atmosphere. Is this how life got started on Earth, out of organic material produced by lightning storms? We don't have the answer yet. And there are other, more surprising places where scientists have found organic material. The meteorites that delivered most of Earth's water—carbonaceous chondrites—also contain organics; a large diversity of organic molecules, like amino acids, sugars, and fatty acids, traveled through space on these ancient messengers.

Before I started searching for life in the cosmos, I just assumed scientists knew how it started on Earth. We don't! This fundamental question is still an area of active research. But we have figured out some of what it needs. First, life needs water. Second, it needs a solid surface where chemicals can stick together and form bonds and structures. These two requirements could be found on the rocky bottom of a pond or on the ocean floor or maybe even on the icy bottom of a puddle on an ice sheet.

We still don't know how the mixture of elements assembled into self-replicating molecules in which tiny variations allowed for improvement over time, leading to all the living world around us. The environment needed to provide energy and the right chemical conditions for these molecules to become encapsulated in a membrane and make a cell. That is how life escaped the ever-diluting power of the oceans and began billions of years of exploration to conquer the world.

When you mix the chemicals for life in water, you need a way to concentrate them so they can start assembling the building blocks of life: RNA and DNA strands and a cell-like structure that can contain them. Scientists are making significant strides in their research on how life got started on Earth. But there is one fundamental problem: scientists can't make life in the lab yet. For now, such a feat belongs to the imagination of visionary writers like Mary Shelley and her creation, Dr. Frankenstein. There are many reasons why making life in the lab is incredibly challenging. How do you set up the experiment (at what temperature, in salty or fresh water)? How long do you need to wait for life to get started (ten minutes, a year, ten thousand years, one million years)?

We know roughly what the Earth's surface was like about three and a half billion years ago—that's when rocks recorded fossils of the first known life-forms. While we know that Earth's surface was warm and covered with liquid water, we don't know if these conditions were what life needed to get started. Maybe life started before a record of it became written in stone. And we don't know if it began nearly everywhere at the same time or if it started in a tiny niche and spread over the whole globe. Life could have begun on the bottom of the ocean in places called white and black smokers, where hot water gushes out of the ocean floor into a freezing, deep part of the ocean under high pressure. Tube-like networks close to such smokers could have provided the surface for pre-cell-like structures to form. Sharp temperature differences can concentrate chemicals and might have been enough to create pre-cell-like and pre-RNA-strand-like structures. If life started like that, it would not have known about sunlight or air. It would not have cared whether the planet's surface was frozen solid or nice and warm. Is the bottom of the ocean where life on Earth started?

Surprisingly, new experiments show that harsh UV radiation—the part of sunlight that can destroy your cells (the reason you need to use sunscreen!)—might help kick-start life. A certain level of UV radiation helps make specific chemical reactions more efficient, which could turn chemicals suspended in water into organic molecules. But UV radiation cannot penetrate the depths of the ocean, so if UV radiation is needed to start life, life must have started closer to Earth's surface. We still need to fulfill the two conditions: water and a surface where chemicals can stick together. Shallow water in ponds, lakes, or ice sheets would provide those. And if some water evaporated, that would concentrate the chemicals, giving them a better chance to meet and stick together. If life started in shallow water, it would have seen sunlight from the beginning. Both ideas of the origin of life are intriguing, but until we can mix the chemicals and make life from scratch in the laboratory, we won't know which one is correct. And even then, it is possible life could have started in more than one way.

Welcome to the world of scientists and how we think. If we see a problem too big to tackle, we pull it apart into smaller, more manageable pieces. We do that until we have pieces of the problem small enough to solve individually. Once we have a smaller solution, we try to fit that specific piece into the frame of the enormous puzzle. How to make life is one of these puzzles. So we take it apart, creating smaller and smaller problems to tackle first, problems like what chemicals do you need to make a cell-like structure? How can you concentrate those chemicals so cell walls form? What chemicals do you need to make RNA? The list of questions and puzzle pieces goes on and on.

Trillions and trillions of chemical reactions must have taken place on a warm, wet, young Earth to lead all the way to life. These transitions might have been inevitable, or they might have been exceedingly rare. We won't know until we have searched other worlds for signs of life. But it happened at least once, here on Earth, leading to you. From tiny microbes to animals, life evolved over billions of years from a single cell to the complex collaboration of trillions of cells that is you. Actually, your body is a fascinating community of human and nonhuman cells: Each of us has about thirty trillion human cells—more than the number of all the stars in our galaxy—but they are outnumbered by the cells of diverse microorganisms that live in and on us. This collaboration is what makes our bodies work. The increase in complexity from the first cell to us is stunning, a testimony to what billions of years of evolution can achieve. It makes me wonder how often a similar process might have started elsewhere in the cosmos.

Every new bit of information we glean from experiments, the rock record, and the new worlds on our cosmic shore is vital to solving this puzzle. The meager rock record of Earth's first billion years provides some insights, but most information about what early Earth was like is lost in time because those ancient rocks have long since been destroyed by the movements of tectonic plates and erosion. Finding signs of life on other worlds can tell us what life in general needs to get started. If we find worlds teeming with life, but those worlds are all frozen, that suggests that life on Earth probably started on ice. If we find no life on frozen worlds but plenty of life in balmy conditions, then warm surfaces are likely what life needs to get started. All of this comes with a warning: we can see only a snapshot in time, how the planet looks *now*, not what conditions were like when life started. But maybe we'll encounter some worlds young enough for our telescopes to glimpse life starting. Even if we don't, if we can catch thousands of rocky worlds at different stages in their evolution, we could decipher the evolution of planets through thousands of snapshots at different ages.

Rock records give clues as to when life gained a foothold in our world, but how long that took is still unclear. After the period of heavy bombardment, Earth's surface cooled and water vapor became liquid; large oceans covered the newly minted surface, and the planet began to resemble the Pale Blue Dot we recognize today. The first small landmasses rose from these enormous oceans. We know for sure that life began to take hold three and a half billion years ago on this eerily strange young Earth. But we cannot say if it existed before then. Finding evidence for earlier life is incredibly hard, mainly because there are few rocks older than three billion years. Most rocks get recycled, subducted or weathered away. To make it even harder for scientists to find clues, only some remnants of life are preserved well as fossils.

In natural-history museums, the displays show more fossilized bones than fossilized skin or feathers. That is because hard bones are much easier to preserve than soft tissue. That is also why the discussion is ongoing as to whether dinosaurs had colorful skin and feathers or not. But before bones and shells could be preserved in the rock record, life had to learn how to build mineralized skeletons and hard shells, something that happened only about five hundred million years ago. It probably took that long because of the high energy cost to make them. Why did it happen then? Scientists believe the rise in oxygen combined with the better chance to survive attacks by predators made it worth the energy expenditure. Fossils began to capture the large diversity of animal life that roamed our planet starting around the time of what is known as the Cambrian explosion. The fossils from the Cambrian explosion tell an intriguing story of the incredible creativity of life.

But even though it's hard to imagine that this epoch was unique in its diversity, very early life, organisms like microbial mats floating in water, had no hard parts to preserve.

Only some of these ancient microbial worlds were petrified and preserved in the rock record. Stromatolites, for example, are fossil reefs built by communities of microbes on ancient seafloors, long before animals took over most reef-building on Earth. Even now, in environments where seafloor microbes are shielded from animals and seaweed, microbes still form stromatolites, revealing a fingerprint of life of ancient times.

Life Gains a Foothold

In 1674 the Dutch scientist Antoni van Leeuwenhoek discovered the astonishing multitude of tiny organisms in a single drop of water, organisms that were hidden away from prying eyes because they were too small to see with the naked eye. Inspired by Robert Hooke's book *Micrographia*, which was published in 1659 and depicted fabulously detailed illustrations of magnified textures from insects to plant life, Van Leeuwenhoek created exquisite compound microscopes, revealing an astonishing world of minute life-forms.

About two hundred years later, in 1859, Charles Darwin's *On the Origin of Species* postulated evolution through natural selection. About one hundred years after that, in 1953, Watson and Crick published the structure of DNA in the scientific journal *Nature*, informed by images taken by British chemist Rosalind Franklin and New Zealand–British biophysicist Maurice Wilkins, opening the gates to understanding life on a new level.

Single-celled organisms were the first life-forms on Earth, and they dominated our planet until about two and a half billion years ago, when life became more complex. A cell is the basic unit of life, and it's bounded by a membrane, a selective wall that allows some molecules and ions in and pushes others out. The cell also stores an instruction manual on how to rebuild itself—its genetic code. We can look at a cell to learn more about life in general. But even these basic units of life on Earth can be incredibly different in size and shape.

Cells can be tiny, with a few thousand of them stretching only about a millimeter, or they can be huge. Individual nerve cells can be up to 4 feet (more than a meter): One reaches from the base of your spine to the tip of your big toe. Probably one of the winners in a heavyweight cell contest, hulking by comparison, is an ostrich egg yolk, about 3 inches (8 centimeters) in diameter and about one pound (half a kg). Knowing this, I have a new relationship with breakfast eggs—the egg yolk, fascinatingly, is only one large cell. Researching this made me wonder about eating an enormous ostrich egg—a taste test of one of the heaviest cells in the name of scientific inquiry.

Probing the surviving rock record of the evolution of Earth's surface and atmosphere sheds light on our planet's incredible changes through time and suggests signs of life to search for. But the oldest samples of Earth's atmosphere—air bubbles trapped in Antarctic ice—are only two million years old. Thus, we have to tease everything we know about a younger Earth out of rocks that formed in contact with ancient air and water.

It would be so much easier if we had a time machine—and if Einstein hadn't shown that a time machine to the past would not work. I would love to take a ride in one to watch a young Earth change. If you are ever offered that ride, there is one critical thing you need to bring. What is it? When I pose that question to my students, I usually get a wide range of answers. Cameras normally rate top. I would love for an imaginary time traveler to take a camera to the young Earth, but if that is the one thing you bring, you would open the time machine's door and—die. Because the chemical makeup of Earth's air today is very different from what it was when Earth was young. The oxygen we breathe is a late addition to the mix, at least in the quantities that allow us to survive. That tremendous change is due to life on our Pale Blue Dot. So if Einstein is proven wrong and you can time-travel to a young Earth, bring an oxygen mask. Then please also bring a camera, a chemistry set, and whatever else you can fit into the time machine to add to our sketchy knowledge of early Earth's history.

Modern Earth's atmosphere is about 78 percent nitrogen, 21 percent oxygen, and 1 percent everything else. But before the rise of oxygen about two billion years ago, the air was made up mostly of nitrogen and carbon dioxide and 1 percent everything else. On a young Earth, there was no oxygen in the air to breathe. Earth's story did not include oxygen until the first organisms that produced it emerged. Oxygen started to build up in our air only about 2.4 billion years ago—shortly after the Earth's two billionth birthday—in what is known today as the *Great Oxidation Event*. Surface rocks before that time contain minerals that are easily destroyed by oxygen. After that time, they don't. The rock record tells the story of a world where life arose and completely changed its planet.

Microbes are limited in their complexity because they get only so much energy. So when cyanobacteria evolved to use sunlight and water as an energy source, a revolution started. They began to split water into hydrogen, which helped power the cell, and generated a waste product—oxygen. This initiated an enormous atmospheric pollution. While oxygen is critical for humans to survive, for most life-forms at that time, it was a disaster. Oxygen produces a large range of reactive atoms and molecules that can damage proteins and DNA, requiring organisms to evolve defense mechanisms. Most single-celled life could not evolve fast enough to survive in the new environment. But for life that learned to use this new energy source, combining oxygen with carbon-rich material, a lot of energy became available, opening up the possibility for multicellular life. Another step closer to us.

Will life on other planets also discover the freely available sunlight as an energy source, develop some kind of photosynthesis, pollute the air with large amounts of the waste product oxygen, then figure out how to use it?

When animals diversified about 540 million years ago, another step in the evolution leading to us, the air on Earth contained up to 10 percent oxygen. But providing energy was not the only novelty oxygen introduced. Oxygen in the air also led to the ozone layer high in Earth's atmosphere, about 15 miles (~ 25 km) up, which protected Earth's surface from the cell-destroying UV radiation. The ozone layer turned the land into a safe enough environment to explore, allowing life to move out of the water. From then on, weary human time travelers could survive on Earth if they'd forgotten to bring their oxygen masks.

Today, plants, algae, and cyanobacteria generate the oxygen we breathe by photosynthesis, a reaction that uses the Sun's energy to turn six carbon dioxide molecules and six water molecules into a molecule of the sugar glucose, one of the essential building blocks of life. This process also produces six oxygen molecules. Two reactions cycle oxygen and carbon between organisms and the environment: photosynthesis takes carbon dioxide, water, and energy and turns it into glucose and oxygen. Then, when organisms eat the plants containing these organic glucose molecules, the glucose reacts with oxygen and produces energy they can use. But that is not the whole story. Organisms in the ocean use carbon to make shells, effectively trapping the carbon there and leaving free oxygen behind.

Oxygen concentrations in the air rose slowly even after the Great Oxidation Event; about two billion years ago, the amount of oxygen in the air was not even 1 percent of what it is today. But not all life on Earth evolved to use oxygen. Anoxygenic organisms offer us a glimpse into the biosphere that dominated our planet before the rise of oxygen. A thriving biosphere on a young Earth-like planet could be made of these life-forms that could not survive in an oxygen-rich atmosphere. On Earth, anoxygenic organisms remained untouched by oxygen in environments such as the sulfur pits in Yellowstone, which present an eye-catching range of colors to visitors. These environments serve as templates for scientists searching for life on worlds devoid of oxygen. We find such niches in many places on Earth, places that life colonized to exploit the resources no one else coveted. Life is surprisingly tenacious.

From biota producing the astonishing hues at Yellowstone National Park to life thriving in the orange-red Río Tinto in Spain (vividly colored due to the river's extreme acidity and very high levels of iron and heavy metals), these extreme organisms show us just a small range of the environments life has conquered here on Earth. And they give us a glimpse of just how different worlds like these could look through our telescopes.

By capturing the colors of these different forms of life in strings of computer code, I can watch them transform my model planets. I can cover the oceans with a green algae bloom or dot continents with yellow microbial mats. After spending a long time searching for any information on how this diverse range of life would look to our telescopes, I built a lab to grow these organisms and collect the necessary information myself. That allows me to encode different life-forms into hundreds of lines of numbers that tell my computer how these organisms interact with light and the atmosphere. Without leaving my office, I can create new worlds.

This is as close to time-traveling as it's possible to get. On my computer screen, a young Earth takes shape. Volcanoes erupt and mark the atmosphere with clouds of toxic gases, then the first wisps of oxygen float through the air, the first thin ozone layer forms to protect the planet's surface, and the first life ventures onto land, adding their color to our world. The sunlight that illuminates our planet shows a beautiful, ever-changing world. The light gets filtered through the atmosphere and sent off into the cosmos. It carries a picture of our world at the time the light started its journey. It also offers a template for how to search for worlds like ours somewhere else.

If you can't travel to other worlds and collect samples to

analyze under your microscope, you have to come up with creative ways to find traces of extraterrestrial biota. Life can change a whole planet, just as life on Earth did. It evolved over billions of years to use the most abundant energy source: light. Through mutations and natural selection, some single-celled organisms developed light receptors and found a way to use light to power reactions in the cell, a major innovation in life's history. Then, about a billion years after that, another innovation occurred that led cyanobacteria to use water, CO_2, and sunlight to produce energy—and the waste product oxygen: photosynthesis. This ushered in an era in which complex life that needed more energy could flourish. Evolution used the higher energy to power more complex mechanisms that eventually led to you and me being able to wonder if we are alone.

Twenty-Four Hours of Earth's History

We know that Earth is four and a half billion years old. To put the billions of years that our planet has existed in perspective, I'll describe the Earth's evolution as if it had taken place over a period of twenty-four hours. Thinking about Earth and life in terms of a full day, from midnight to midnight, provides startling insights into how monumentally Earth has changed through its evolution. Humans are intimately familiar with only a very short, precious, and fleeting moment in our planet's history— the time that Earth's climate has allowed you and me and our fellow humans to strive and wonder. In the grand scheme of Earth's evolution to date, humankind is just a short blip on the four and a half billion years of our planet's existence. It is up to us to stretch that time into the future.

Imagine Earth formed twenty-four hours ago at midnight, 12:00 a.m. The rock records show that life had begun by 5:00 a.m. (3.5 billion years ago). That start time could have been even earlier, but there's no record to track it. Oxygen rose and transformed our air starting a little before lunchtime (about 2.4 billion years ago). Fossils of the first multicellular life date to 1:00 p.m. (about 2.1 billion years ago). First land plants developed at about 8:00 p.m. (750 million years ago). The Cambrian explosion occurred around 9 p.m. (530 million years ago), and it marks the nearly simultaneous emergence of myriad diverse animals with hard shells. Plants became widespread and began to color the landmasses green at about 9:30 p.m. (450 million years ago). Oxygen reached a concentration of 15 percent around 10 p.m. (400 million years ago), letting a human time traveler breathe unaided for the first time since the planet formed. The mighty dinosaurs roamed on Earth for an hour, between 10:40 p.m. and 11:40 p.m. (250 to 66 million years ago). The Himalayas started to rise about 11:45 p.m. (50 million years ago). *Homo sapiens* stepped onto the cosmic stage a few seconds before midnight (about 300,000 years ago), and the first radio signals leaked from our planet a fraction of a second before midnight (100 years ago).

What if you were a time traveler without a cosmic star map—would you be able to recognize Earth on your trip? In the 1968 sci-fi movie *Planet of the Apes*, based on the novel by Pierre Boulle, human explorers land on a planet and recognize that it is Earth only when they find the Statue of Liberty buried in the sand. But imagine you were one of these explorers. Could you recognize your own world? What would you use to identify it?

Let's look at the recent changes first, the ones we are familiar with. *Homo sapiens* appeared about 300,000 years ago, a few seconds before midnight. All the landmarks we use to orient ourselves have been around for only a blink of an eye in geological time. The Himalayas and the Alps have been around for only a few million years, since about 11:45 p.m.. Continents crashed together and broke apart again and again through Earth's history, creating mountain ranges we can't even imagine, let alone identify on a younger Earth. Those mountain ranges won't provide you with any known landmarks. Even dinosaurs started to roam the Earth only about 250 million years ago, around 10:40 p.m. Trilobites populated the Earth from 530 million years ago, around 9 p.m. to about the era of the dinosaurs, but would you recognize your planet based on the trilobites swimming in the oceans? Or would those strange Earth creatures look like alien life-forms? Even the widespread green land plants are missing until about 450 million years ago, around 9:30 p.m.; before they appeared, the land was barren, and the scenery would have been unrecognizable to a modern human. Even the sky was devoid of familiar star constellations, providing no comfort.

If the dial on the time machine returned to the new day's first hours, the young Earth you landed on would appear to be a completely alien world.

Colors of Earth

Any hypothetical extraterrestrial astronomers pointing their telescopes toward Earth would see our planet at a time in our past, the farther away the younger our planet would appear. To paint the colors of our world through time, you need an artist's color palette.

Black Earth: A young Earth, with its cooling black magma crust, muggy, steamy atmosphere, and a black alien Moon looming huge in the sky, looked like a world out of a sci-fi movie. Mega-volcanoes released vast amounts of toxic gases into the air, creating a world shrouded in hazes and clouds.

Blue Earth: Once Earth cooled down and oceans engulfed it, our planet started to look like a blue dot from space. The young Moon cooled too, and our large gray companion initiated massive tides circling this young world; here and there, the first landmasses started to rise from the endless seas—a surfers' paradise, if you didn't need to breathe. Over billions of years, more and more continents rose from the oceans, creating a gray pattern among the blue seas.

Red Earth: After two and a half billion years, Earth's oxygen started to accumulate in the air. The land was still completely barren, and with the increase of free oxygen, the land began to rust. Oxygen reacted with the surface minerals, turning parts of the Earth red. For a while, Earth may have looked like a large, wet Mars, with red continents separated by blue oceans.

White Earth: Earth has frozen over several times in its

geological history; the evidence was captured in the rock record 2.4 billion years ago and a second time 650 million years ago. This temporarily changed our planet into a white, crusty, ice-and-slush ball enveloped from poles to equator in ice. Scientists think it got cold because the ongoing massive outpouring of volcanoes led to rapid loss of CO_2 from the air around the warm, fresh volcanic rock, reducing the greenhouse effect and initiating worldwide glaciation. Once the planet was frozen, these weathering processes slowed. Still, volcanic outgassing of CO_2 kept going, eventually reaching levels that set off immense warming and catastrophic melting, transforming Earth into a hot world that then turned temperate.

Pale Blue Dot: When life started to conquer the landmasses, around 450 million years ago, Earth changed again, displaying colorful green continents painted by the biosphere embedded in blue ocean. Our planet's vast oceans, its medley of white clouds, and the scattering of blue light in our atmosphere combine to paint our modern world an astonishingly beautiful, fragile, pale blue. A brilliant speck on the vast black canvas of space. Carl Sagan's words echo in my mind whenever I think about our place in the cosmos: "Consider again that dot. That's here. That's home. That's us."

We know life can change a planet because it did on Earth. Earth's history chronicles its profound transformation and allows a first glimpse of the diversity of one rocky world and the life it shelters, hinting at what to look for.

SIZE COMPARISON OF THE SUN & EARTH

PLACES TO SEARCH FOR LIFE IN THE SOLAR SYSTEM

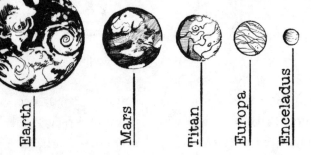

Earth

Mars

Titan

Europa

Enceladus

Chapter 4

How to Search for Life in the Cosmos

I have loved the stars too fondly to be fearful of the night.

—Sarah Williams

Colors of My World

In Upstate New York, seasons change in a flurry of colors. The fresh green dotted with pink and white spring flowers gives way to the yellow cornfields of summer, then to the vibrant orange leaves in autumn, then to the austere white scenery of winter, and then it starts all over. Most noticeable is the green that's everywhere most of the year. If you shine a light on any surface, part of the light will be reflected back. Vegetation appears green because it reflects green light.

Most oxygen-producing plants on Earth use green-pigmented chlorophyll, which means they reflect green light but use red and violet light to power photosynthesis. Almost half of Earth's photosynthesis is done by the four hundred thousand species of land plants that tint our continents a beautiful green. But on a younger Earth, life was different. Alternative pigments have evolved through Earth's history—algae, bacteria, and lichen in a wide range of colors have harvested solar energy. While it would be comforting to find green trees on planets circling other stars too, as you have seen on Earth, green vegetation became widespread on land only about 450 million years ago, which means if you were an alien looking for green vegetation only, you would be blind to life's signatures for most of the time it existed on our planet. A wide range of colors provide a key to finding life on Earth. We won't be able to step foot on any exoplanet for a long time, but we can catch light from such planets and moons in our telescopes. Could other worlds look the same as ours?

When you enter my lab, the first things you notice—after the seemingly crazily arranged flasks and containers everywhere—are the beautiful swaths of color. Each petri dish is home to a specific kind of life that fills it with yellow, green, or red hues. Algae paint their homes a beautiful pink, red, orange, and green. Just imagine a new world with an ocean covered in red algae blooms, its seas looking fiery red.

Why am I, an astronomer, in a biology lab growing different kinds of microorganisms? It is detective work. How do you catch thieves? By looking for the fingerprints they leave behind. Life has its own fingerprints. It can cover a world in multiple shades. It would be a pity for scientists to miss signs of life in the cosmos because they were looking only for green plants. But there was no database for the colors of life that someone looking through a telescope could see. So I decided to create a color catalog of life in all its amazing shades, spanning as many species as I could find. In this sense, *find* meant talking my biology colleagues into giving me samples they had collected for various other purposes. The samples represent biota from many places: sunbaked dry deserts, ice sheets in the frozen Arctic, hot sulfur springs, and outside your door. It is easy to forget how incredibly varied life on Earth is. Organisms adapt to survive in their habitats, and the results are vastly different lifeforms, from jellyfish to stick insects. However, microbes alone take this diversity to an astonishing level.

But while growing microorganisms is easier than growing, say, animals, it is not as easy as it might sound, especially for an astronomer. When I was a student at a summer program at Johns Hopkins University in Baltimore, I grew brain cells in the lab, and when I moved my cells to a new container, I promptly killed 96 percent of them. That is apparently normal for beginners. The university invited me to come back the next year anyway, which I took as a good sign, but by the next summer, I had discovered exoplanets and the search for life in the universe, and my heart was taken. Understanding how to grow life in the lab comes in handy now on my quest to find it on other worlds. So, how not to kill these diverse biota samples?

First, I convinced the microbiology and remote-sensing departments that they really wanted to have an astronomer in their labs. That is why I built the interdisciplinary Carl

Sagan Institute at Cornell—because I knew astronomers could not find life in the cosmos alone. We have the most amazing resources in our colleagues from other departments, although, admittedly, they're sometimes quite puzzled as they try to figure out what exactly the astronomer wants. But this network of scientists and enthusiasts has led to unique and incredible collaborations. One of these collaborations allows me to scout for colorful organisms around the globe and catch their light fingerprints, their cosmic portraits, so we can identify them if they appear in our telescopes. And I greatly appreciate my amazing team of astrobiologists who can grow diverse forms of life without killing most of them, because I know how hard it is.

Once the organisms are grown, we need to learn how they would appear through a telescope. My colleagues and I pile the dozens of vials filled with colorful organisms into a backpack (a low-cost transportation device) and walk across campus to the remote-sensing lab in the civil-engineering department. In this second lab, we set up a spectrometer with a funny-looking sphere that resembles a gold-plated Magic 8 Ball and put the different organisms into the sphere so the light will hit our sample from every direction, just as it would on the surface of a planet.

Then we measure the light that bounces off our sample. Our spectrograph, the instrument that analyzes the light in shades of every color, tells us exactly how light bounces back off the surface. We take that information—the fingerprint of reflected light—and put it in our toolkit so if there are, for instance, huge algae blooms in pink, orange, red, or green on another world, we will know what to look for.

Searching for Life: Where Do We Start?

We are only one of many species in a long line of life on our beautiful planet. Life on Earth is breathtakingly varied. Imagine the bottom of the ocean and the deep-water life there, organisms with intriguing names like sea angels, crossota jellies, pink see-through fantasia, deepstaria, dumbo octopus, and vampire squids. They stun and inspire me with their bizarre beauty and diversity. And the deep oceans are just one of countless places life calls home. Eight million species of animals and plants live on our planet, and that's a conservative estimate. If you could scan a one-page description of each in one minute, you would need more than fifteen years—without taking a break—to read all of them.

The vast diversity of life on our own planet can be sorted in four main categories based on its source of carbon and energy. Life can take carbon directly from its environment, the way plants draw CO_2 from the atmosphere or from water—these are autotrophs (*auto* means "self," and *troph* is "to nourish"). Life can get carbon from consuming organic compounds, which is what animals and humans do—these are heterotrophs (*hetero* means "different"). Life can obtain energy through sunlight (*photo*), which is what plants do, or from chemical reactions that break down organic and inorganic compounds (*chemo*), which is what animals and humans do. Humans are chemoheterotrophs—we acquire both energy and carbon from food. Plants are photoautotrophs—they get their energy from sunlight and their carbon from CO_2. Some bacteria are photoheterotrophs—they obtain energy from sunlight and carbon from food. But there is life on Earth that needs neither light nor food—chemoautotrophs. Some bacteria draw their energy from chemical reactions and procure their carbon from CO_2, which means they can inhabit environments that you and I could not survive in.

There are some ultimate limits for life, though: temperatures and pressures under which structures—cells, DNA, and proteins—will break down no matter how well protected they are. But within these limits, we find life thriving on Earth in a huge variety of environments. It's found around white and black smokers, natural chimneys of sulfide minerals that pump out very hot water around 600 °F (~ 300 °C) that's black with iron sulfide into the frigid, dark ocean; these support a thriving community more than a mile (two km) deep in water with high concentrations of hydrogen sulfide. In slushy brine in the Antarctic, single-celled algae harvest energy from the sunlight filtered through ice and assimilate nutrients from the water below. Even in hot sulfur springs and soda lakes, life grows vigorously. But despite its seeming diversity, all life on Earth is built from just twenty-four elements of the more than one hundred that scientists know of. All proteins in living organisms are built from the same twenty-two amino acids, although there are more than five hundred other possibilities. And all life that we know uses DNA (or RNA). If life were to evolve somewhere else, would other recipes also work? We don't know if these compounds won out on Earth due to chance or if there is something critical in these particular combinations. It's an active field of research. Finding alien life that's built with other amino acids and other solvents would provide an answer.

Where should we start looking? Is there a hard limit on the color of starlight we need? It would be easy to say we need to look for yellow suns. Earth has one. But I hope, after this discussion of the different categories of life, you can shed that Earth-centric worldview. What if a planet's sun is a different color? Let's take a red star. Vegetation looks green because it reflects back the light it does not use to generate energy. If the light suddenly changed, Earth vegetation would be in trouble; the green basil happily growing and releasing an enticing smell on my kitchen counter would likely struggle and die if I took it to a planet bathed in red light (which carries less energy than yellow light). But that is because plants evolved photosynthesis here on Earth, under the light of a yellow sun. If a process like photosynthesis evolved on another world, it would use the color of that starlight to produce energy. Plants are green under a yellow sun, but they would likely be darker under a red sun to catch all the incoming energy. That might result in gothic-looking black vegetation, but life should flourish even under a red sun. Even on Earth, we find organisms that can use red light to gain energy. Why? There are communities of organisms that live together in microbial mats. Some of the organisms are on the surface and have access to yellow sunlight, but some are in deeper layers of the mats that the yellow sunlight does not reach—those organisms evolved to effectively use the leftover light, meaning the red light. And, as we have seen, some life does not even need sunlight. Those life-forms would be fairly nonchalant about changes in colors and light conditions.

For me, imagining what life on other planets could look like is creative and inventive, but I wonder if our imagination can cover even a fraction of the possibilities. When I see the newest images of deep-sea creatures, I must admit that I could not have predicted their astonishing, weird, strange beauty beforehand. If we find life elsewhere, its appearance will be an additional, beautiful aspect, full of surprises.

Searching for Life Does Not Mean Searching for Us

The astonishingly vivid colors of the sulfur springs of Yellowstone National Park make up for the smell of rotten eggs that assaults your nose when you get too close. The gorgeous array of colors painting the landscape makes it a unique place on Earth, drawing millions of visitors annually. I visited Yellowstone for the first time in 2008 for a meeting of NASA's Astrobiology Institute. I had completed my Ph.D. only a few years earlier, but I was sent there to speak for our whole team because all the team leads were on other trips. It was both exciting and intimidating to present our research to the leaders of other teams. But it was a great opportunity for me to learn from the top experts in their fields in an astonishing location. When you visit Yellowstone with a group of astrobiologists who study life in extreme conditions—which colors the ponds in Yellowstone everywhere—you can point to any place as an example of the peculiar, weird, incredible capability of life.

An orange-colored rim on a hot sulfur spring is more than just beautiful—it comes alive with stories of the organisms that color it, who found them, how those organisms grow, what they can and can't do, and the adventures of collecting them.

●

I arrived late at night. I had opted for one of the cheapest flights that got me to an airport close to the park. My flight landed around ten p.m., and it was fully dark outside. I'd come from Boston, and I had not realized how dark it would be out here. It created an astonishing sky-watching opportunity but also made it hard to see any potential obstacles in the road—such as roaming elk. I had not thought of Yellowstone and elk together until I read the email from our organizers warning us that it was elk-mating season. Naturally, I had rented the cheapest, tiniest car I could find, smaller even than one of the roaming elk.

My car was the only one on the pitch-black road to Yellowstone. I was inching along so I could brake if an elk ventured onto the road. Admittedly, I had no idea what I would actually do if an elk took an interest in my tiny car. The email from the meeting's organizers had advised visitors not to get between an elk and its mate, not even in a car. This was not particularly useful advice if you didn't know the location of the elks.

Back at the airport, the car-rental agent had asked where I was going and then inquired twice if I was sure I wanted to decline the accident insurance. I did decline—company guidelines. That slow drive under the light of thousands of stars as I scouted the deep darkness for large moving shapes is etched in my mind as one of the adventures of being a scientist: finding my way with the assistance of the light of the stars (and Google Maps), venturing into the wild (okay, not really the wild—it was a lodge at Yellowstone).

Finally, after several long hours, I arrived at the lodge. The attendant at the front desk was highly amused by the combination of the size of my car and the late hour.

I hadn't seen any elks on the road, but the next day, I woke up to the eerie calls of elk cutting through the predawn. Elk-mating rituals were an impressive spectacle, but they could not compare to the stunning natural spectacle of Yellowstone National Park, with its hot sulfur springs rimmed in beautiful colors, their hues changing from blue to green to red and yellow. Most of these colors—except for the deep blue in the middle—indicate various organisms that thrive in an environment that would kill most life.

Watching these colorful displays made me notice how broad the range of colors of life are. Just imagine a planet with hot sulfur springs covering its surface, where life creates a rainbow-colored world. It was then I realized that astronomers needed a color catalog of life—a database of diverse Earth biota and information on how they reflected incoming starlight—to compare to what our telescope would find on exoplanets, leading to the range of vials of biota in different shades in my lab. Our colorful comparison chart that included more than just green plants.

On Earth, most of these biota exist in beautiful but limited niches. But on other worlds, such conditions might be normal, and life could evolve—as it did here on Earth—for those conditions. Maybe these imagined widespread hot sulfur springs would lead to multicellular life-forms perfectly adapted to those worlds.

Extremophiles are what we call organisms that live in extreme environments. These life-forms are generally small microorganisms that thrive in extreme conditions. Extremophiles are adapted to conditions humans would not survive. And extreme conditions are plentiful, even on Earth. But extreme is in the eye of the beholder. If extremophiles could talk, what would they say about humans? They would probably lament the terrible conditions we have to endure—so cold compared to their hot sulfuric springs, our water lacking the tasty acidity, and so on. *Who could survive that?* they might think. Whatever you evolve for, that is your normal.

So the term *habitats* includes conditions that are not survivable for you and me. Actually, it is sobering to realize that for most of Earth's history, humans would not have been able to survive on this planet. If we could rewind time and start again, it seems unlikely that Earth would produce humans again. A planet with different starting conditions and paths of evolution has no obligation to support life similar to Earth's, let alone curious humans.

Let's pick another one of the millions of life-forms on Earth, one outside the limelight, and see what its habitable conditions are like. Meet the tiniest hero—you have probably stepped on one. The tardigrade looks like it would be at home in a sci-fi movie, although it's only a fraction of a millimeter tall. Often called water bears or moss piglets, tardigrades (the name means "slow stepper," a reference to the micro-animal's slow movements) can be found almost anywhere there is liquid water, from altitudes over 19,000 feet

(6 km) in the Himalayan mountains to ocean depths of more than 15,000 feet (4.5 km) from tropical rain forests to beneath layers of solid ice. Tardigrades have four pairs of legs that end in either claws or suction disks. A microscope reveals these explorers.

Tardigrades have survived all five mass extinctions on Earth, of which the demise of the dinosaurs was the last. They live for about two years, but that's not counting their dormant state. Unlike us, they can survive for hundreds of years in a dormant state, then wake up and go happily about their lives again. Some tardigrades can survive temperatures down to −450 °F (−272 °C), very close to absolute zero, and up to 300 °F (~ 150 °C). You can cook or freeze them; they won't mind too much! They can survive pressures more than a thousand times that found on Earth's surface, radiation hundreds of times higher than the lethal dose for humans, and the vacuum of outer space.

In 2016, I gave a talk on my research at the annual astronaut conference, the Association of Space Explorers Congress, held in Vienna that year, and it was amazing to meet people who had seen Earth from space—not just in photos but in reality. Several of them told me that seeing the endless, nearly desolate darkness of space made them feel a completely new connection to our planet. Earth's atmosphere distorts the image of the stars, which is what makes them appear to twinkle. Once you leave the Earth's protection, stars don't twinkle anymore, they shine steadily, and you find yourself in the complete silence of space, alone except for a glimpse of the Pale Blue Dot that is home.

In my talk at the conference about the search for life

on these new worlds, I might have slightly misjudged my audience when I referred to tardigrades as the perfect astronauts, lifting the tiny water bear over human astronauts' collective incredible achievement of leaving Earth. But the actual astronauts did not hold it against me (at least not for long). My point was that tardigrades in space needed neither protective gear nor food, which made space travel much easier. You could put tardigrades on a flimsy spaceship with no life support, fly them to Mars without feeding them on the way, then sprinkle a few drops of water on them, and they'd be ready to go. Try doing that to an astronaut. (I did not actually suggest that.) Unfortunately, we have no way of communicating with a tardigrade; we can't tell it what to do or teach it to drive a Mars rover. So astronauts are much better than tardigrades in space if you want to do more than just save money on survival gear.

Tardigrades look like they could be from another world, and in a way, they are. Their world is a drop of water, a microcosmos we can't visit except for brief periods when we see it under a microscope.

Water bears are actually not extremophiles; they are not adapted to exploit and thrive in extreme conditions, only to endure them. Tardigrades can suspend their metabolism and survive without food or water for up to one hundred years, sometimes more. I learned that when I sat next to the director of the National Museum of Natural History in Paris at a conference and excitedly told him about tardigrades. He told me they had just sprinkled water on a two-hundred-fifty-year-old tardigrade in stasis, and it happily woke up and walked away. See, that is why going to conferences is important. It's about all the people you talk to randomly and learn things from, in addition to listening to talks on the newest research. But as the French team has not published its findings yet, you'll hear this only as an anecdote. The published time tardigrades can survive in their dormant stage is one hundred years. Tardigrades dehydrate before they go into stasis. This apparently otherworldly survival mechanism is the answer to a very earthly problem—surviving water shortage. When water dries up, so do the tardigrades. They shrivel up into balls, lower their metabolisms, and continue in a kind of stasis until they encounter water again, at which point they carry on as if nothing has happened. This capability makes them incredibly space-resistant.

In 2007, a European research team decided to test that resilience. They sent three thousand tardigrades into orbit around the Earth for twelve days—on the *outside* of a rocket. The project's name was Tardigrades in Space, TARDIS for short, an acronym that has a special meaning for fans of the famous BBC sci-fi series *Doctor Who*. On *Doctor Who* TARDIS stands for Time and Relative Dimensions in Space and is the name of the time machine of the Time Lord, Doctor Who. Project TARDIS was a success: Tardigrades are the first animals that survived exposure to outer space, which would kill a human in about ninety seconds. The TARDIS experiment was not the last time tardigrades became space travelers; some were unsuspecting passengers on a one-way mission to the Moon. We'll talk about that a bit later.

Searching for Life Close to Home

Places in our solar system where life could thrive might look very different from the beautiful green landscape around us.

Our nearest planetary neighbors are Venus and Mars. Venus is blanketed in a thick CO_2 atmosphere and shrouded in sulfuric acid clouds, so not a good contender for the search for life. If you look away from the Sun, you'll see our neighboring planet Mars, which provides a much more promising—albeit freezing-cold—environment. Mars is smaller, and its gravity is only about one-third of Earth's. The red planet has a much thinner atmosphere than Earth because its gravitational embrace is weaker.

An atmosphere is made of atoms and molecules that move faster the hotter the temperature is. Gravity can keep gases from escaping into space—from reaching escape velocity—and leaving the planet or moon behind. The story of lost gases is painted on Mars's soil. Mars is red because most of its near-surface water is gone. Water was split into its components, hydrogen and oxygen, and the lighter hydrogen escaped Mars's low gravity and went off into space. The little bit of oxygen produced in this process colored the surface red—basically, Mars's surface rusted. But Mars's color is only skin-deep, a fraction of an inch. Scratching the surface reveals the region that oxygen did not touch—Martian soil is actually light brown.

Because Mars has a lot less atmosphere than Earth, it can't stabilize the temperature between the day and the night side. So unlike on Earth, where days and nights hold similar temperatures, daytime temperatures on Mars can be up to a balmy 70 °F (~ 20 °C), while nighttime lows go down to −230 °F (−150 °C), as we saw earlier.

Sandstorms on Mars can engulf much of the red planet for a long time. And every five years or so, Mars gets blanketed in a planet-wide sandstorm that can last for weeks. But storms there are different from storms here on Earth. In Mars's thin atmosphere, even a giant storm's one-hundred-mile-an-hour winds would feel like a gentle breeze.

Storms on Mars would not topple spacecraft or any structures we might build there in the future. But fictional storms can make for great stories. In one of the first scenes of Andy Weir's highly enjoyable novel *The Martian*, a giant sandstorm nearly topples a spacecraft. That would be impossible. Andy Weir knowingly fudged that detail, but this scene artfully sets the intriguing survival story in motion.

About four billion years ago, Mars might have been a warm blue planet covered with oceans and rivers like ours, and it might have had all the ingredients for life. But unlike Earth, Mars is so small that its internal heat was not enough to keep its core molten and moving. As we have seen, once the innards of a planet freeze, its surface does not get pushed under anymore, so tectonics and volcanism stopped replenishing gases in Mars's atmosphere about 3.5 billion years ago. Without climate cycles, Mars cooled further and further. The feeble light of a young Sun reaching this planet's orbit beyond Earth could not push its surface temperature above freezing again. Mars's atmosphere is mainly CO_2, an effective greenhouse gas, but there is not much of it, so Mars does not get much greenhouse warming. Water on Mars today is mostly locked in ice and permafrost at the poles. Maybe liquid water survives underground, but it can't survive on Mars's surface anymore. Today the red planet is a freezing-cold desert but the surface pressure is so low that liquid water boils away in seconds. Mars also does not have a radiation shield like the magnetic field and the ozone layer of our Pale Blue Dot. So if there was ever life on Mars, or if there is today, microbes are the strongest contenders hiding under the soil to escape the radiation. That is why missions to Mars are trying to find traces of life underground.

Mars has taught us a profound lesson: habitability can be temporary.

Journeying farther out, the giant planets in our solar system inhabit the freezing realm far beyond Mars's orbit and are reached by only a trickle of sunlight. That dim sunlight can't support rivers or oceans of water on the surface, so any water that exists is locked below frigid ice sheets. Especially two small rocky moons covered in ice offer intriguing destinations to search for life. Each circles a majestic giant planet. If you stood on the icy surface of Europa, you would have a stunning close-up view of Jupiter; on Enceladus you'd be treated to astonishing views of Saturn, the beautiful giant with brilliant rings. The icy crusts on these moons hide potentially habitable oceans.

That these moons harbor oceans came as a staggering surprise to astronomers.

Both frigid moons are small. Europa is about the size of our Moon, and Enceladus is even smaller, only about 300 miles (~ 500 km) across, about half the length of the state of California. Large sheets of crusty ice cover the moons' surfaces, but they show cracks that would not be there if the whole moons were frozen solid. Remember that water is less dense when it freezes, so ice floats on subsurface oceans. Gravity, that tireless architect of the cosmos, turns out to be the key to this mystery. These moons do not circle their planets alone. Both icy moons get stretched and compressed by gravity, kneaded like bread dough due to the gravitational pull of some of the other moons and their planets. That energy keeps the water below the thick ice sheets from freezing just as dough gets warm when you knead it vigorously. These thick, crack-covered, icy crusts could provide shelter for life hidden in dark oceans.

Europa is one of Jupiter's biggest moons. It could hold more than twice the amount of water in all of Earth's oceans. The chilly average surface temperature on Europa never rises above −260 °F (−160 °C) at the equator. The moon was discovered by Galileo Galilei in 1610 and caused quite a stir because Galileo saw through his telescope that the four Galilean moons—Io, Europa, Ganymede, and Callisto—circle Jupiter. Through a telescope, you can watch these four small but bright dots of light, just like Galileo did hundreds of years ago. (Jupiter's moons are named after four of the Roman god's lovers, in case you want a romantic reason to look at the sky on Valentine's Day.)

That the Galilean moons do not circle Earth, proved that not all bodies revolve around Earth, putting an accepted theory on shaky ground. Jupiter's moons have already helped reshape our worldview by revealing that the Earth is not the center of the universe. Can they do it again by providing another habitable world in our solar system?

We don't know yet. But in 2023, ESA's mission JUICE (an acronym for Jupiter Icy Moons Explorer) blasted off and will arrive at Jupiter in 2031, followed by NASA's Europa Clipper mission in 2024 that is scheduled to arrive in 2030, the different flight times due to the alternate gravitational dances they perform along the way.

Saturn and its moon Enceladus are even farther from the Sun, so the average temperature on the surface of Enceladus is an astonishing −330 °F (−200 °C). Yet there are liquid oceans deep below the frigid surface of this tiny moon as well. The NASA-ESA Cassini-Huygens mission shared breathtaking views of Saturn and its icy moons for more than a decade and plunged through the rings to analyze them. And the mission discovered the oceans under the sheets of ice on Enceladus.

This tiny moon spews jets of water high above its surface, creating a beautiful spectacle; a plume of water particles that freeze and glitter in the chilling darkness of space. Those frozen pieces can tell us what these oceans are made of.

The Cassini spacecraft flew through the plumes of Enceladus and analyzed their composition, with intriguing results: the water contained some organic material, indicating that the ocean on Enceladus might just be able to host life. But scientists have no way to learn the deeper secrets of these geysers—yet. Plans to send a probe loaded with instruments to sample the plumes of Enceladus for signs of life are already on the drawing boards of space agencies worldwide.

Icy moons in our solar system have many mysteries, some of which we are about to unveil.

A Place for Life as We Don't Know It

There is one other place in the solar system covered by oceans and streams, but they are not made of water.

That place is a moon covered with an orange haze. It's surface temperatures are much too low for water to flow. While water freezes because it is so cold, other chemicals stay liquid even at these extremely low temperatures. The sunlight that reaches Saturn's moon Titan is about one hundred times fainter than Earth's, leaving the moon with a harsh surface temperature of −290 °F (−180 °C). But Titan is brimming with organic material. Methane and ethane carve river channels and fill great lakes and seas, sculpting its surface. Even a hidden water ocean might flow far below Titan's frozen ground.

Saturn—and Titan—are nearly 900 million miles (~ 1.5 billion km) from the Sun. Titan is tidally locked in synchronous rotation with Saturn, always showing the same face to its planet, like our Moon with Earth. A night on Titan is a bit shorter than on the Moon, only about 8 Earth-days long. It takes Saturn 29 Earth years to circle the Sun, a Saturnian or Titanian year. I would still be waiting for my second birthday party on Titan, but seeing a brilliant Saturn in the hazy sky might be worth missing a few birthdays.

The mission carried a passenger to the Saturn system: the Huygens probe, the first human-made object to land on a world in the outer solar system. In 2005, it plummeted through the dense, hazy atmosphere all the way to the surface, recording what it saw, and providing a new picture of this intriguing moon.

Large parts of Titan are covered by dark hydrocarbons that look a bit like dunes made of coffee grounds. Titan's hydrocarbon seas are named after mythological sea monsters; you could plunge into the Kraken Mare there. The names of Titan's mountains come from a different mythology, the fictional world of Middle-Earth created by J. R. R. Tolkien. If you were on Titan, you could hike up the Moria Montes (the Mountains of Moria). Walking on Titan's surface would feel like moving around on the seafloor about 50 feet (15 meters) deep in one of Earth's oceans.

Titan is larger than Enceladus—it's even larger than the planet Mercury—but it only has half the mass of Mercury, creating a very low gravitational pull on its surface, less than that of our Moon. So future space explorers will be able to jump even higher on Titan than the astronauts did on the Moon. But the cold air on Titan is thick, about one and a half times as dense as on Earth. So if you strapped wings to your arms, you could fly on Titan. You could not breathe, but you could soar high on this −300 °F hazy moon.

NASA is planning to fly an eight-bladed rotorcraft on Titan soon: the Dragonfly mission with a launch date in 2027 should arrive there in 2034. Like a drone it will loft its scientific payload to a dozen different sites on Titan in order to explore the surface and look for organic material.

This hazy moon is an intriguing contender for a habitable world, one very unlike our own.

Spacecraft and landers are already scouting the surface of Mars with increasingly sophisticated rovers and helicopters that can analyze soil samples for microbes and fossils. Such exploration carries with it a tiny but potent danger: unknowingly brought-along hitchhikers. If we find life in another place in our solar system, we'll first have to ask ourselves if we brought it there. It is incredibly challenging to kill all life on a spacecraft or rover before you send it on its way, so if we find life on Mars that looks similar to Earth's life, how will we know if it's a native Martian or an Earthly hitchhiker? We wouldn't know for sure unless it was an entirely different kind of life, down to the equivalent of an alternative form of DNA. But it's not only spacecraft stowaways that could make life we find on neighboring worlds similar to ours. All the bodies in our solar system have shared material since the beginning. When an asteroid collides with a planet or moon, rocks escape the gravity of that world and are sent out into space. Mars gets hit by meteorites more than a hundred times a year, and over a thousand tiny meteorites come crashing down on Earth every year. Most of the material falls into the ocean or on uninhabited land, so luckily—or unfortunately—the chances that a space rock will crash down in your garden are small.

What if these rocks carried their own hitchhikers?

Microbes on bigger meteorites, at least one meter in diameter, might survive the trip between planets and moons. Meteorites resemble brilliant fireballs as they fall to Earth, heated by our atmosphere, but the interior could stay cool enough for microbes to survive. Theoretically. Even though there has been some debate about possible findings indicating life on Mars, including U.S. president Bill Clinton's 1996 announcement of just such a finding in a Martian meteorite, unfortunately we have not found any Martian life-forms yet. The idea that microbes could travel between planets (a concept called *panspermia*) fuels the imagination. And if life can hitchhike on meteorites, then we could all be Martians, seeds of life brought from the red planet to flourish on Earth. That idea is intriguing, but when you compare Mars and Earth in terms of our criteria for a habitat (energy and liquid water on a solid surface), Earth wins. Mars could have sustained surface water for only a brief amount of time compared to our Pale Blue Dot.

Given that, Earth is the likelier bet for where life started. So we are probably not Martians, but maybe Martians, if we ever find them, are really *ancient Earthlings*. For now, if you want to call yourself a Martian, go ahead; no one can prove you wrong. Hopefully, if we find life on Mars, we will find that it developed independently from Earth life, making it a genuine alien. If it looks similar, we'll wonder.

Is There Life on the Moon?

There is one other body in our solar system with life: the Moon. It traveled hundred thousands of miles to get there. A private Israeli lander, Beresheet, brought an experiment containing thousands of tardigrades, the miniature voyaging heroes we met earlier, to the Moon. But disaster struck, and the mission—and thousands of tardigrades—crash-landed. The tardigrades were part of a project of the Arch Mission Foundation, the mission statement of which is to "archive humanity's heritage for future generations." Another small test library, a quartz disk containing Isaac

Asimov's Foundation trilogy, had already been launched into space in 2018 in the glove compartment of a Tesla sports car launched as the dummy payload in the Falcon Heavy test flight. The sports car is currently circling the Sun once about every eighteen months as an artificial satellite.

To distribute more of these archives throughout the solar system, another DVD-sized time capsule of humanity was included on Beresheet's lunar lander. This time capsule consisted of thirty million pages of information as high-resolution, nanoscale images engraved into nickel. The first four layers held nearly all of the English Wikipedia and thousands of classic books, as well as the key to decode the remaining twenty-one layers. And between the twenty-five nickel layers, each only a few micrometers thick, were layers of epoxy resin, the synthetic equivalent of the tree resin that preserves ancient insects; it held human DNA samples—and thousands of dehydrated tardigrades in their dormant stage. The tardigrades were a secret addition to the design.

As we have seen, in their dormant stage, tardigrades can survive a lot: You can boil them, freeze them, dry them out, or send them into space. The tardigrades might have even survived the crash landing. In 2021, scientists in the U.K. ran tests to see if the tardigrades could have survived the impact. First, they froze them for forty-eight hours, which lowered their metabolism by nearly 100 percent and sent them into their dormant tun state, a state of suspended animation. Then they placed them in hollow nylon bullets and shot them into a sand target at higher and higher speeds. Tardigrades survived impacts of up to 2,000 miles

(~ 3,000 km) an hour. At speeds higher than that, the scientists reported, they were "just mush."

The last measurements received showed that the Beresheet lander was traveling at 300 miles (~ 500 km) per hour, but it is unclear what the final impact speed was. Did its Earthly passengers survive? The only way to know is to go and search the wreckage on the Moon.

Should people be allowed to just litter the Moon with tardigrades? Space agencies have a protocol in place, developed at NASA's Planetary Protection Office and others, to avoid contaminating places in the solar system that could host life. But the Moon did not make the list of possible habitats, so it did not get this protection.

Surprisingly, spilled tardigrades were not the beginning of the pileup of organic matter on the Moon. Astronauts have left about one hundred bags of excrement, in addition to cameras, boots, and other things so they could lighten the load for the return trip, meaning that future humans or alien visitors will be able to get insights into the human diet and some of our history by analyzing the content of those trash bags. Leaving their garbage behind allowed the astronauts to take back more Moon rocks. Scientists on Earth got more samples to study, and any alien archaeologists will have plenty of materials to evaluate in the future. Although they will probably be puzzled by how those bags got to the Moon and where the organisms that produced the waste went.

If, millions of years from now, any aliens found thousands of dormant tardigrades on the Moon, would they think the tardigrades were the species that managed to leave the Earth's gravitational pull behind? That the tiny water bears were explorers scouting our solar system? When I look up at the Moon, I imagine the dormant tardigrades lying there like thousands of Sleeping Beauties, suspended in time until they are brought back to life, not by a kiss, but by a few drops of water.

Under a Purple Sky

Our sky is blue only because of the chemical makeup of our air—it scatters blue light better than red light. Light pings off molecules and particles in the air like a ball in a pinball machine. High-energy light—the bluer part—pings off more particles and is sent scattering through the air in all directions. Lower-energy light—red light—pings off fewer particles and moves forward less hindered. That means more blue light than red hits your eye, and when you look at the sky, it seems to come from everywhere. Except at sunset. When the Sun sets at the horizon, there is more air between you and the Sun than there is when the Sun is high up in the sky. The pinball effect of scattering still works, and the blue light gets scattered so much between the horizon and you that more red light reaches your eyes. That is why the sky looks red at sunset.

But the color of our sky can change too. Large wildfires, like those in 2023 in Canada or in 2020 around San Francisco, can change our atmosphere's composition just enough—in that case, by adding dust and soot particles that scattered red light as well. This addition made the sky look orange, like in an apocalyptic sci-fi movie. That begs the question: What colors can a sky be? It turns out it depends on what the air is made of and if there are particles like dust in it. Air on other planets could have a very different chemical makeup than ours, so the pinball effect of scattering would be different too. Pick a color to paint an alien sky. Imagine a pink sky or a purple sunset. It might exist on one of these new worlds. I would love to see that, but first, I'd make sure I was in a safe place on a spaceship or on an enclosed base, because a sky painted in eerie colors means that breathing the air would probably kill you. So beware, future astronauts, of different-colored skies! Certainly, some blue skies could kill you too—a younger Earth had a blue sky but no oxygen to breathe yet—so this advice is not bulletproof.

And not every world gets a colored sky. Photos on the lunar surface always display a black sky. The Moon's gravitational embrace is too feeble to hold on to much gas. Without an atmosphere light does not get scattered, so the sky looks black.

Everyone knows the beautiful nursery rhyme "Twinkle, Twinkle, Little Star"—but it is our air that makes the twinkling. As we have seen, stars don't actually twinkle. Hot and cold airstreams in our atmosphere distort the image of a star before it hits our eyes. So, eerily, if you're on the Moon or any world without an atmosphere, the sky will be black and the stars won't twinkle—the parents of future babies on the Moon will have to adjust their nursery rhymes.

Finding Life over Cosmic Distances

But if there is an atmosphere on another world, it could tell us if life already exists there.

Light and matter interact; starlight can make molecules vibrate and rotate. Every wavelength of light carries a unique energy, and to make a molecule move, the light must have just the right amount of energy, no more and no less.

Imagine two different molecules, oxygen and water: O_2 is made up of two oxygen atoms linked together, H_2O is made up of three atoms: two hydrogen atoms and one oxygen atom linked together. Due to the way atoms bind, each molecule has a unique structure. You need a specific push and energy to make oxygen swing and another to make water swing.

Light encounters molecules and atoms on its way through the cosmos, and when it reaches us, the missing parts of the light reveal what chemicals it encountered en route. The missing parts—scientists call them *spectral features*—are a bit like passport stamps that tell you which countries someone traveled through before they arrived here. But unlike passport stamps, the missing light provides clues to the chemical makeup of the air on the worlds it passed through. To decipher those clues, scientists measure which specific energies (that is, which light colors) make which molecules move or electrons jump energy levels in an atom. But you cannot measure everything. Scientists use calculations instead of experiments when the required temperatures would melt or freeze the laboratory equipment or the gas mix would poison the experimenters.

When I finished my dual degree in engineering and astronomy, I was hired to optimize the design of a space telescope that could spot life light-years away. The space telescope was called Darwin, and it was one of the candidates selected by ESA, which envisioned a fleet of telescopes searching for signs of life on exoplanets (in the U.S., there was a similar candidate, called the Terrestrial Planet Finder). Darwin would allow scientists to probe the atmosphere of Earth-like exoplanets for the first time.

But there was a glaring problem: no one knew how many of those existed and how many stars a telescope needed to survey to find them. So our design team made some educated guesses. If every tenth star had a planet that was like Earth, and Darwin wanted to characterize three such planets, then the telescope needed to survey at least thirty stars. If every hundredth star had an Earth-like planet, then Darwin needed to survey at least three hundred stars. NASA's Kepler mission would tell how many potential Earths there were out there, but this was 2001, and that Kepler mission wouldn't be launched for another eight years. Our best guess was that one in ten stars would host an Earth-like planet, and we designed Darwin for that.

But there was another issue that bugged me. *Modern* Earth was the standard used to design a spacecraft to search for life. But Earth had changed since its earliest days, and the chemical makeup of its atmosphere had as well, so I was sure that Earth's spectra, its light fingerprint, must have changed too. Science was missing a huge piece of the puzzle—an understanding of how Earth's light fingerprint had changed over time. Without that, any telescope was likely to miss the signs of life unless the planet was a carbon copy of modern Earth.

A colleague of mine once told me that everything you have experienced in life shapes how you see the world, what you think is important to do next, and how you approach solving problems. I tried to convince many scientists that we needed to create a model of what Earth's spectra would have looked like from the moment the planet was formed until today (and even beyond). But that turned out to be an extremely complicated task, not just because it required answers to many connected questions but also because it required input from several fields—geology, biology, astronomy, and engineering. And interdisciplinary science, which is when scientists from different specialties work together, was only in its infancy. Long gone are the days when scientists could know *all* the scientific knowledge of their time. That is great news because it means that humans have discovered much more than any single person could ever learn, and every day, new information is added to our database of knowledge. But it also means you cannot do everything alone anymore.

In science, we stand on the shoulders of giants—all the people who came before us and figured things out. They provided us with a shortcut to that knowledge so we don't have to read all the articles and redo all the experiments to come to the same conclusions. But there is a downside to all this knowledge too. Finding answers by bringing people from different specialties together is a big challenge—not because scientists don't want to work with one another, but because each field developed independently and has different critical goals. Even something as simple as the meaning of words varies from field to field. A geologist will talk about large differences in the composition of the atmosphere and mean one part in a million; an astronomer will talk about large differences in the distances to stars and mean trillions of miles. Imagine them talking to each other without specifying exactly what they mean.

To model the light fingerprint of Earth through time, you need answers to many questions. How has the Sun evolved since Earth formed? How has the chemical makeup of Earth's atmosphere varied? When did continents appear? When and how did life change the planet's air and surface? Every one of these questions could fill the lifetime of a scientist. And once you have these answers—and many are still under debate—you need to create a computer program that can produce the light fingerprint of a faraway planet, bringing together all you have learned.

This was a critical problem, and yet no one was tackling it. But that changed in 2001 after I had a somewhat random encounter with a scientist at a conference. I was talking about the importance of creating the spectra of our planet through its evolution so we would not miss signs of life if it was in a different evolutionary stage than us. Stars are not all the same age, I said, so the same must go for their planets. The American astronomer Wesley Traub, then at the Harvard-Smithsonian Center for Astrophysics in Boston, told me, "If you really think this is worth doing, you will have to do it yourself." To change people's worldview, you need to show them the importance of what they can't see yet.

I sometimes think it was extremely fortunate that I had no

idea how hard it would be to get all the answers I needed—if I'd known, I might never have tried!

To generate the computer code that could explore a planet's environment, I needed to connect with and learn the worldview of many scientists from different departments, so that's what I did. Most of them were quite surprised at the barrage of questions from the astronomer in their midst. I needed to connect the ideas of people who had different training, insights, and skills to address the complex problems inherent in searching the cosmos for life.

It took me three years to create the model that generated the first light fingerprint of Earth's evolution through time. Along the way, I learned how much we don't know about our planet's evolution, such as how, when, and where life started; Earth's evolution and its critical importance in helping us find worlds like ours has fascinated me ever since. With the fossil record, we can extrapolate what Earth was like when it was younger, although there is greater and greater uncertainty the further back in time you go. Using that information to determine what kind of place a younger Earth was, I created a computer model that generated these light fingerprints: the prebiotic atmosphere of a young Earth, with large amounts of CO_2 wrapping like a blanket around our planet to warm its surface, creates a different light fingerprint than the atmosphere when dinosaurs roamed breathing oxygen. These light fingerprints changed significantly as our planet matured and life left its imprint on the air and surface.

The results of my years of work were intriguing. For about half of Earth's history—around two billion years—

our atmosphere showed the telltale signs of life in its light fingerprint: oxygen together with a reducing gas like methane (reducing gases are compounds that react with atmospheric oxygen). Spotting individual life-supporting elements like oxygen is not sufficient proof of life. On a hot world where intense radiation breaks up water into its two components, oxygen and hydrogen, oxygen will be produced in large amounts and could fool a hopeful explorer. We have to be careful in interpreting what we see, especially because we are *hoping* to find signs of life. That is why scientists are trained early to ruthlessly kill their darlings. By carefully modeling what a wide range of Earth-like worlds with and without life would look like to our telescopes, we can work out which signs can be trusted and which should make us worry. That way, we can flag troublesome candidates before we start celebrating. And it allows us to identify the conditions that cannot be explained by anything other than life. For now, our best bet is an atmosphere with a combination of oxygen and methane on a planet that is in the habitable zone of its star. So that Goldilocks region is where astronomers point their telescopes.

Signs of life on Earth are a glass-half-empty-or-half-full kind of situation. I could have discovered that life on Earth has been detectable for only a few hundred thousand years, and that would have made our search for life on other planets much harder, so two billion years is much better. But I could have also found that Earth's light fingerprint has shown signs of life ever since life left a fossil record to decipher, about three and a half billion years ago. That would have made the search much easier because signs of life would have been detectable earlier. But the fact that Earth's light fingerprint has shown signs of life for two billion years already gives us a long window of time to find it in another world. But given that scientists know that a younger Earth will show weaker oxygen features, they know how much longer to point their telescopes at younger planets for a long time to catch even small wisps of it.

My results also made me wonder if an alien astronomer could have spotted life on Earth already, since our light fingerprint has shown a thriving biosphere for two billion years already—but more on that later.

Our solar system hosts a multitude of planets and moons, and I became curious about what their light fingerprints were like. The light fingerprint of Mars is different from say that of Venus, or Jupiter. But different how? To answer that my team created a light-fingerprint catalog—something like the database of fingerprints police uses to match prints from a crime scene—for the most diverse planets and moons in our solar system: majestic Jupiter, frigid Europa, dazzling ringed Saturn, cool red Mars, hellish Venus, and our Moon, among others. The Solar System Spectra Database contains the light fingerprints of nineteen bodies in our system and serves as a basis for comparison for our search out there in the cosmos. Finding another Earth would be breathtaking, but finding another Mars would also be intriguing. And what about a super-Enceladus or a mini-Saturn? The Solar System Spectra Database allows us to compare faraway worlds with the neighbors where landers and rovers can roam. But so far, the exoplanets we have already found are not copies of the planets in our own solar system. To interpret what we are finding, we need to push the boundaries and envision types of planets that we don't have in our own system, planets that emerge from a combination of strings of letters and numbers.

My computer screen shows my austere code, scrolling white letters and numbers on a dark background. Thousands of lines together create a program that reveals how the energy hitting the atmosphere on a world circling another sun changes it. Temperatures rise or fall, chemical reactions destroy or produce gases, heat is captured or released—a new world appears in front of my eyes. A few keystrokes let me move the planet closer to the star, manipulate the color of its sun, heighten its gravity, create worldwide sand dunes, oceans, or jungles, and add or remove life-forms. I am creating worlds that could be and the light fingerprints to search for them with our telescopes.

Life can change a planet. But while an alien observer would have been able to find the signs of life, oxygen with methane, in Earth's atmosphere for about two billion years—dashing the hopes of humans who want to keep life on Earth a secret—that scientist would have no way of knowing what kind of life was here.

That's another intriguing component to the search—if we find signs of life, we'll have no idea what kind of life we've found. It could be anything from microbes to plants to animals that use oxygen; there is no straightforward way to know. Once we find signs of life, that will be our next adventure—figuring out what kind of life we are looking at.

Biofluorescent Worlds and Glowing Aliens

If life exists on the surface of a planet under a red sun, it must be able to withstand harsh conditions. Some small red suns shower their planets with flares, bursts of high-energy UV radiation, especially when those stars are young. High-energy UV radiation can destroy cells and genetic material and consequently can be used to sterilize medical instruments—the last time I went to my dentist, I noticed that they sterilize their equipment with UV radiation.

When I started to imagine and model habitats under a red sun, I became really curious about how life on Earth protects itself from UV radiation. Here, different life-forms have different strategies for this: some repair the damage; others shelter from UV radiation underground or in water; others generate defense mechanisms, like pigments. We humans put on sunscreen. Remember that on Earth nowadays, not a lot of high-energy UV radiation makes it to the ground. So humans did not have to evolve ways to deal with extreme UV radiation.

Before the Great Oxidation Event changed Earth's air and generated the ozone layer that protects Earth's surface, life thrived in the oceans where it was sheltered. Water absorbs harsh UV radiation, providing protection. But surprisingly, even in the ocean, some life-forms developed an effective and beautiful way to combat high UV radiation: they glow.

This glow is due to biofluorescence, and it's different from bioluminescence, which is when light is generated by chemical reactions in an organism's body for specific goals. Think of a garden full of fireflies, they use bioluminescence to communicate with one another. If you live on the coast of Puerto Rico, Jamaica, Vietnam, Japan, or the Maldives, bioluminescence sometimes makes the ocean glow bright neon blue-green. The water is brimming with dinoflagellates, single-celled plankton with tails that have made some of Earth's coastlines glow for more than a billion years.

Biofluorescence is something else. It's what makes some life-forms glow under the light of a blacklight lamp that emits UV light. Next time you dive into the deep ocean, take a blacklight with you—some of the fish and corals you shine that light on will glow. Biofluorescence allows organisms to absorb high-energy light waves and then re-emit them in a glow of brilliant fluorescent blues, greens, pinks, oranges, and reds. A surprising number of species glow under UV light—fungi, plants, and even animals. A wide range of fish, seahorses, salamanders, frogs, puffins, scorpions, opossums, and owls glow. Chinese fire belly newts glow a fiery orange, platypuses glow purple-green, wombats neon blue, and flying squirrels a beautiful pink. Scientists are still learning why some life glows under UV radiation. Theories range from using it as a means to communicate to protecting symbiotic organisms by breaking down harsh UV to harmless visible light.

You can see biofluorescent corals in aquariums worldwide. In 2020, following a lecture I gave on the search for life in the cosmos at the California Academy of Sciences in San Francisco, I convinced one of the museum curators to take me down to the aquarium after hours. (As a scientist, I get to meet many interesting people, like curators who have the keycards to visit museums when these are closed to the public.) All the lights were off. In the velvet darkness, I could just make out a soft glow. As I stood alone in front of a huge glass window, I got a glimpse into this beautiful glowing underwater world that left me spellbound.

Imagine a world out there circling a red sun that throws out violent UV flares. When the radiation reaches the planet, the oceans could light up with a soft colorful glow as life breaks down the harsh energy, protecting itself in a beautiful spectacle. Standing in the dark, I imagined what such an incredible display of evolution could look like.

THOUSANDS OF NEW WORLDS DISCOVERED

Chapter 5

Worlds That Shook Science

Two roads diverged in a wood, and I— I took the one less traveled by, And that has made all the difference.

—Robert Frost

A Winding Road

Since the start of my career, I have been propelled by my fascination with the question of whether we could ever find life on other planets—and how I might be able to play a role in that exciting research. But the road to scientific discovery is not without its potholes, especially for women. Sometimes they are big enough to block your path completely, requiring creative maneuvering to find a new way to follow your dream. (I can recall numerous incidents where I was challenged or ignored, experiences that will resonate with many others, no matter what gender, and might provide a little help when facing these obstacles.)

"This is just crazy," my Ph.D. student Sarah declared, outraged, upon bursting into my office. She was upset on my behalf because she overheard two men on the bus stating with conviction that I had been given my position as the leader of one of the highly competitive Emmy Noether research teams here at the prestigious Max Planck Institute for Astronomy, only "because I was a woman." I'd been awarded one of the sought-after grants while I was a researcher at Harvard University, and that was why I decided to come back to Europe. You might think comments like this could be taken as a joke, but they are never said jokingly. Such criticisms are even leveled at women who have been awarded Nobel Prizes in the sciences, the highest honor in their fields.

The fact that the program, funded by the German Research Foundation to the tune of about half a million euros, is named after the famous German *female* mathematician Emmy Noether was probably lost on these men too. A few months earlier one of my male colleagues had taken it upon himself to announce to no one in particular at one of the first coffee meetings at my new workplace that I had gotten "the women grant." But although named for a woman, the Emmy Noether grant is awarded to only a small number of "exceptionally qualified early career researchers"—both male and female—from any science discipline after a rigorous selection process based on merit and vision. If your application makes it through the first tough selection, the members of a large scientific committee drill you further about any practical challenges or tiny flaws they see in your proposed research in an hour-long in-person interview. The committee's job is rightly to ensure that the program's substantial investment is well spent. So how did I respond to that male colleague at coffee? I innocently and loudly asked him if in German all awards named after *male* scientists are just for men. It made most of the other scientists present smile.

This incident made me realize that my feelings were surprisingly complicated at that moment. I was proud and grateful for Sarah's support, sad that my colleagues felt comfortable making such pronouncements about one of the institute's few senior female scientists on a public bus for all to hear—but I realized that I was no longer surprised at such comments. I had, in fact, become somewhat resigned to them when they concerned me, but Sarah's outrage on my behalf shifted the tone to a hopeful note: it had never crossed her mind that anyone could be anything other than outraged by the comment she'd heard. Sarah, an amazing scientist in her own right and now a professor of astronomy, had been my Ph.D. student at Harvard, and she was visiting me for a few months so we could finish a project together. She is part of a new generation that gives me hope that things will get better, that we are creating a place where everyone belongs in science.

I can recall many prior incidents when the role of women in science, including my own, had been challenged. In one job interview for a top scientific position, one of the first questions I was asked was if I had children, apparently more relevant than my experience or vision for the job, which the committee had yet to discuss. I answered "yes." It was hardly a secret. Next question: "And you are married?" Having established that I had a child they now were apparently curious how that had come about. I was not sure where this line of questioning was going. If I was not married—I happen to be—then would my interrogator be looking for details about the identity of the baby's father? Two sides can play this game, and I decided on a bit of humor. Am I married? "Yes," I responded, "to my laptop!" What hiring committee doesn't love a workaholic? Another panel member hastily jumped in before the concerned colleague could utter the logical follow-up questions if I was pregnant or planned to be anytime in the future.

In situations like this, I cling to a German proverb passed on by an older friend that has no good English translation: "Don't get upset; instead, wonder why." (*Nicht äergen, nur wundern*). While it might be counterintuitive, this practice causes me to focus on other people's motivations, to question how people arrive at their view of the world. Switching my perspective helps me to not get too frustrated—at least for a while. But even though not many scientists are still trapped in the Stone Age in terms of their views of the female half of the population, some of the ones who are can have enough power to block someone's path in competitive environments. How many brilliant ideas and discoveries have already been lost because gifted young women had to use most of their energy to fight to be even allowed to do research?

Throughout the course of this book, I hope to convey just how difficult the search for alien life will be—to the extent we might not even recognize it when it is staring us in the face. The very best chance for humankind to be successful in this epic quest is to have the broadest, the most diverse, spectrum of thinkers working together. People from all backgrounds, cultures, and genders are needed, in the hope that in pooling our different perspectives we will broaden the scope of our thinking and expertise sufficiently to make the breakthroughs we need.

So, stepping in to keep things fair and ensure a level playing field is crucial to make planet- and alien-hunting, as well as science in general, a prosperous place for everyone. It is sometimes as simple as to speak up, and politely state that you don't agree. Silence makes the speakers think everyone agrees with them, even if most people actually don't. They just think it is not their place to speak up. Not staying silent is a crucial piece of breaking this vicious cycle.

I'd managed to avoid most of these detractors by chance, really, which might explain how a girl from a small town in Austria is now searching for life in the cosmos with the best international teams.

When I took a test in high school to evaluate the studies I was best suited for, the examiner suggested that I stay away from the natural sciences because that was not really something women could do. (Unsurprisingly, there are not many women cited for scientific breakthroughs in history because they could not even enter the fields, let alone are credited with the discoveries themselves if they managed that first hurdle.) My parents were outraged, successfully torpedoing the examiner's advice. It is critical to have someone remind you that you can learn whatever you want. For me it was my parents who kept reminding me of this when

I was younger, and later more and more friends and colleagues encouraged me. For anyone who is struggling to find their place, I suggest that you find one person who provides that kind of encouragement for you and then hold on to their advice.

At the university I attended, I encountered this Stone Age vision less than I expected but more than I had hoped. Faculty members held mixed views on whether women belonged in engineering and physics and astronomy, but only one, an older professor in engineering, went so far as to ignore the two female students in his class completely, welcoming only the male students at the beginning of each class and joking about women. (He shared in the first lesson the wisdom that physics could not be a random mistress; it could only be the lawful wife, although, of course, that was not to say it wasn't a good idea to have mistresses in real life.) I figured he was just a relic from ancient times, and I was sure that his kind would die out eventually, preferably before I took the final exam. That proved to be youthful optimism, but it got me through the class. And I had many other professors who did not share his outdated opinions, giving me hope. Some were even excited to have a female student in their class, meaning I always had to be prepared, because I was the one student whose name these professors remembered early on and enthusiastically called on.

In my first job out of university, I learned a profound lesson from my boss, one that helped me navigate my work life. I had just been chosen by the European Space Agency to work on the design of the Darwin space mission to search for life in the cosmos. At twenty-three, I was by far the youngest member on the team and the only woman in the future-mission department. But my international colleagues there treated me as an equal. The first week, we had a meeting with a big engineering company. My boss, Anders Karlsson, from Sweden, had prepared a talk with about fifty slides, and as we were about to enter the meeting room, he realized he had forgotten to make copies of the slides for all the participants. I offered to go do that while he got the meeting started. I still remember how he turned around and looked at me to ensure I was paying attention. "Lisa, if you go and copy the slides, you will forever be the secretary and nothing you can do will make them see you as an engineer in your own right." We went into the room, and he told everyone he had forgotten to make copies of the slides and that he would do it right now and we'd start ten minutes late. As I sat in my seat among my male colleagues, it hit me that Anders had just made a pointed statement to everyone about my role on the team. His not accepting my offer to copy the slides seemed like a small thing, but it had a huge impact. And it taught me to pay attention and intervene when necessary, because seemingly small acts can have big repercussions.

I sometimes get asked what career advice I would give to my younger self. I'd tell her, "Develop selective hearing as fast as you can. Find a group of people you trust and listen to their advice. And try not to listen to people who have not earned your respect." I choose to remember not the ridiculous men on the bus but the heartwarmingly courageous, furious young scientist in my office who made it clear to me that being a woman in science is getting better. There's still a ways to go—it was sometimes difficult for me, and I was part of the majority culture in the wealthy environment of Western Europe who had the full support of her family and access to free education. I know how much more difficult the path has been for others without such advantages. But it is getting better. Some of the Stone Age scientists have died out, and there are more women in senior positions all the time who demonstrate that pursuing science at the edge of knowledge should be open to everyone, even those who don't conform to the stereotypes. And many of my male colleagues are part of that process too—they don't just ignore supposedly funny jokes about mistresses but make sure that conferences aren't dominated by male speakers, which can easily happen when you populate the panels and lectures by thinking about which of the colleagues with whom you recently had a beer would give a great talk. And then there is the younger generation, scientists who are already used to working with diverse teams and who have learned that by drawing from a range of experiences they can solve problems faster.

Whenever I still encounter a fellow scientist who challenges me, not because of my work but because of who I am, I'm reminded of another incredibly helpful piece of advice once shared by a senior colleague: "Look at it this way: if someone is trying to take you down, that means you have done things worth noticing." And, like their peers, women scientists have been among those doing many things worth noticing—including getting ever closer to answering the riddle of the presence of life in the universe.

The Planet That Could Not Exist

The blue hues of giant storms intertwined with light gray rivers of air cover most of the planet's visible surface, pushing against each other. The merciless sun heats the winds to speeds higher than any tornado on Earth.

The massive gas expands in the blazing heat and is stripped from the planet by the intense light. Even the gigantic planet's gravity is no match for the speed of the hot molecules hurtling into the numbing darkness of space.

The harsh stellar wind pummeling the planet tears the gas from the heated outer layers. The planetary exodus creates a stunning light show—a comet-like tail of gas flares out from the planet. The tornadoes rage in a losing battle against the stellar inferno. And the planet loses itself little by little, disappearing into the profound darkness of space.

The discovery of new worlds outside our solar system started with a mystery: a tiny wobble. In 1995 two Swiss astronomers, Michel Mayor and Didier Queloz, whom we met earlier, detected a weird signal from the star 51 Pegasi. The star, a near twin to our own Sun, about fifty light-years away from Earth, unexpectedly wobbled back and forth on its stellar journey. And stars don't wobble for no reason.

Majestic Jupiter, the biggest planet in our solar system, contains the vast majority of the material left over from our Sun's creation and provided the first clues as to what afflicted 51 Pegasi. Jupiter makes our Sun wobble just a tiny bit. Jupiter is a humongous ball of swirling gas around a rocky core a dozen Earths big. This colossal gas giant, the fifth planet from the Sun, out beyond Mars, is a sight to behold: stunning patterns of storms cover the whole planet. Monstrous weather systems stir and twist the gases and paint the planet in patterns that look like van Gogh's *The Starry Night*.

If Jupiter were an empty box, all the other planets could fit into it and there'd still be room to spare. Jupiter dwarfs the Earth; you would need to put seventy Earths, one next to the other, to make a belt for Jupiter's middle (a fresh idea for a Halloween costume!). Powerful wind speeds exceeding 400 miles (~ 600 km) per hour create some of the largest storms in the solar system. One of them, Jupiter's *Great Red Spot*, has been observed for over a century—and is large enough to easily engulf Earth. Voyager 1—the spacecraft carrying the Golden Record out of the solar system—sent back the first detailed images of this gigantic storm in 1979.

But compared to the Sun, Jupiter is a lightweight. If Jupiter were a tablespoon of water, the Sun would be a four-gallon jug. If you had a cosmic set of scales, you would need about a thousand tablespoons of water on one side (a pile of Jupiters) to balance the Sun on the other side. In this comparison, Earth would be the size of a waterdrop. To balance the Sun on these cosmic scales, you'd need to place about three hundred thousand waterdrops (a humongous amount of Earths) on the other side. All the planets in our solar system together would make the cosmic scale tilt only a minuscule bit. The Sun is just so massive. The disk surrounding a nascent star contains only a tiny part of the material that creates the star at its center, and that disk forms all of its planets.

It would take about a hundred Earths to span the diameter of the Sun. To imagine this, line up one hundred peppercorns on the floor. (Pro tip: it is extremely helpful if the peppercorns are not the same color as the floor. In my first trial, I used black peppercorns on a dark floor, which, in hindsight, was not the smartest choice.) The one-hundred-peppercorn line shows the vast size of the Sun compared to our Pale Blue Dot. You would need more than one million Earths to fill the inside of the Sun (volume is proportional to radius cubed).

So finding an exoplanet in the vast cosmos is extremely hard. If you wanted to find a planet circling another star, what kind would be the easiest to locate? Astronomers looked around our solar system and chose as their prototype the biggest, most massive planet to look for somewhere else: Jupiter.

The Sun's gravity loses some of its pull at the colossal giant planet's distance, so Jupiter does not need to travel as fast as the Earth to counter its gravitational pull. The balance between gravity and speed determines how long it takes for a planet to complete a circle around its star. While Earth does it in one year, Jupiter takes a leisurely eleven Earth years to circle the Sun. Knowing they would have a slightly easier time finding massive planets like Jupiter than finding tiny Earths, astronomers settled in for a decade-long search.

Like the other giant planets in our solar system, Jupiter

consists mostly of gas and ice because it formed far enough away from the hot Sun that ice and gas did not evaporate, leaving massive amounts of planet-building material as we have seen. It's cold beyond the ice line. Jupiter receives only one photon for every twenty-five photons Earth gets.

Planets are different than stars in more ways than just their size. They do not have nuclear-fusion reactors in their cores, so they do not produce energy and they do not shine. Like our Moon, they just reflect the starlight that hits them. That makes planets tiny, dim objects, incredibly hard to spot beside a huge, illuminating star. Seen from space, the Sun is more than one billion times brighter than Earth to our eyes. Think of it this way: one billion seconds is about thirty-one and a half years. If we compare the numbers not in brightness but in time, you would have to wait more than thirty-one and a half years of starlight to get one second of light from the planet. The light of an Earth is drowned out by the light of its host star.

But planet hunters have clever ways of finding their quarry. When you look up at night, you can see thousands of stars moving across the sky. In most cases, what appears to be the motion of the stars is actually the result of the Earth rotating around its axis and circling the Sun. But sometimes, there is an additional, unexpected motion, an indication that we've spotted something truly spectacular. Because even lightweight planets tug—just a little—on their heavyweight hosts. Both the star and its planet counter the other's gravitational pull by adding a little extra to their movements. Because the star is so much more massive, it wobbles only slightly if a planetary companion tugs on it. But that tiny wobble makes all the difference. It led astronomers to discover the first new worlds on our cosmic shore.

Take a Walk for Science

Imagine someone walking her dog in the park. If the dog and the dog walker don't agree on the direction they want to go, the dog pulls while the person leans back to keep her balance. Think of the leash as gravity binding them together. The bigger the dog, the more the owner needs to pull to determine their direction. If you were to observe this outing from another corner of the park with some bushes in the way, you wouldn't need to see the dog to know that the owner was being dragged in a particular direction by a determined force. Similarly, astronomers can see a star move when a planet pulls on its host. Now imagine the dog running circles around the owner. You would see the owner wobble back and forth rapidly to remain upright. This motion is how astronomers detected the first exoplanets around other suns even though they remain hidden from curious eyes.

When you look up at the stars at night, you can't see them wobble. The tiny gravitational pull of a planet is incredibly difficult to measure. But there is a trick to finding out if a star wobbles. Patterns in a star's light create a precise measuring tool. Every hot object glows in a wide range of colors. It sends out characteristic radiation in the well-understood shape of a blackbody curve. The light is brightest at a very specific color—what that color is depends on the temperature of the object—and it dims fast on either side. Did you ever wonder how astronomers know how hot the Sun's surface is? A nearly unfathomably hot 10,000 °F (~ 5,500 °C)? No thermometer can measure the Sun's surface temperature, and even if there were one that could, we'd have no way of getting it to the Sun because anything we could send would melt at those temperatures. So we need another way to measure the temperature of hot objects like stars. The blackbody curve solves the problem. The color of a radiating object tells scientists how hot it is, so starlight serves as a cosmic thermometer. For instance, we know that red stars have cooler surfaces than yellow stars like our Sun do.

But not all of a star's energy escapes. Stars have thin, scorching gas layers on their surface. We can tell the chemical makeup of this scalding-hot gas by its light. More precisely, we can tell its chemical makeup from the light that is *missing*. This thin surface layer catches some of the outgoing energy, because light and matter interact as we saw.

Every element has a distinct structure, its own unique note in the grand cosmic composition. The electrons within an element's atom only occupy specific energy levels—imagine a stadium, in which you can move from row 1 to row 2 or from row 1 to row 4, but not to row 2.5. Electrons can jump between different energy levels if light bumps into them with just the right energy. But only very specific colors of light have just the right energy to make an electron jump to a different level. So the electrons of different atoms absorb specific colors of light. This interaction creates patterns, kind of like a barcode unique to each chemical element. These patterns tell you which chemicals the light encountered on its path. And this barcode gives us a precise handle on the slightest movement of a star.

You can identify the patterns for different chemicals in the starlight, but the barcode is not always where it is supposed to be. Sometimes the barcode has shifted to redder colors; sometimes it has shifted to a bluer, shorter wavelength (the color red has a longer wavelength than blue). And sometimes, when you look long enough at the same star, the barcode shows an intriguing behavior: it redshifts (appears at redder wavelengths), then returns to where it is supposed to be, then blueshifts (appears at bluer wavelengths), then goes back to where it is supposed to be, then it redshifts again, and so on. The change of the position of the barcode compared to the lab measurements we make tells us that the object we are looking at is moving back and forth—wobbling. And the question becomes: What can make a star wobble?

When you find the right pattern at the wrong color (or wavelength), you see the Doppler effect in action. You have encountered the Doppler effect before: when an ambulance drives toward or away from you, the sound of its blaring siren changes. Or, as Formula 1 fans know, when a racing car speeds by you, the sound of its engine changes. That happens because you are standing still while the vehicle moves relative to you. The pressure waves in the air pile up when the ambulance or race car approaches you, then stretch out when it moves away from you. But the Doppler effect is not limited to sound. It also modulates the vast electromagnetic-wavelength range that includes, among other phenomena, light you can see and infrared heat radiation you can feel. Christian Doppler, an Austrian physicist, realized that the observed frequency (or color) of waves depends on the relative speed of the source and the observer; he published his findings in 1842. In addition to the Doppler effect, Doppler con(Ef)fect(ion), very tasty champagne-vanilla truffles created in his hometown of Salzburg, is named after him—a different kind of timeless memorial.

When a star moves away from you, the barcode pattern gets redshifted, which means that you find the correct pattern at redder—longer—wavelengths than you do in the laboratory. Understanding this fundamental concept led to a long string of breakthroughs. For example, that is how the American astronomer Edwin Hubble figured out in 1929 that the universe was expanding: the barcode shifted. But which way? Do galaxies have redder or bluer barcodes than they should have? Most galaxies in the cosmos, other than the nearby galaxies we travel along with, show redshifted barcodes. So, on average, galaxies move away from us, which is lucky since it means they're not about to come crashing into us. Seeing the barcodes shift to redder and redder colors for galaxies farther and farther away shows us the universe expanding right in front of our eyes.

But what does redshifted and then blueshifted starlight mean? Redshift: the star is moving away from you. Blueshift: the star is moving toward you. A red-blue-red-blueshifted pattern means that the star moved away from you, then toward you, then away again, and so on. This wobble indicates that something is tugging on the star, just as the dog tugged on its owner as it ran around them in the park.

How much the star wobbles depends on how massive the planet that tugs on it is. Gigantic planets with stronger tugs are generally the easiest for astronomers to find.

In a cosmic symphony light conveys the tales of distant stars, galaxies, and planets, their temperatures, compositions, and motions with each color or wavelength. When we explore the range of wavelengths, the spectrum, we're not just observing phenomena, we're listening to the melodious narratives of the cosmos, beautifully written in the language of light.

I described the first exoplanet found in the introduction to this chapter. 51 Pegasi b tugs strongly on its star, 51 Pegasi. Think of the star's name as the family name of the planet and the letter after it as its given name. But 51 Pegasi b was not where it was supposed to be. Astronomers had settled in for decades of searching, but they found 51 Pegasi b surprisingly fast. This huge Jupiter-like planet circles close to its scorching-hot star. It does not touch the star, but you could fit only four stars between 51 Pegasi and its planet. For comparison, you could fit forty Suns between Mercury and the Sun, and one hundred Suns between Earth and the Sun, meaning that 51 Pegasi b is ten times closer to its star than Mercury is to our Sun.

In life and in science, we take what we know (in this case, our own Jupiter) and act accordingly. But life and nature often surprise us. Mayor and Queloz knew that their instrument could only detect massive planets like Jupiter, a planet that creates a wobble signature (the red-blue-red-blue shift) over the course of eleven years. Astronomers can identify a planet when they have observed a bit more than half its path around its star, but that still meant several years of searching if they were going to identify another Jupiter. But they found a star wobbling back and forth in a mere four and a half days. Even Mercury, the closest planet to our Sun, needs about three months to finish a full circle around the Sun. This new world was much faster. A full trip around its sun lasted only four and a half Earth days. Monday morning to Friday afternoon, and the year was over on 51 Pegasi b. This closeness to its sun left the planet sizzling, way too hot for liquid water. In fact, it is so scorching that the heat is boiling away its atmosphere. Part of the planet's outer layer is getting ripped away and is being hurled into the cold darkness of space, creating a sparkling tail of gas that turns to ice in the freezing vacuum.

A hot giant planet so close to its sun that the star strips part of its outer layer—51 Pegasi b was not like anything astronomers had expected. Imagine, for a moment, you were one of the astronomers who found this signal. It made no sense for such a huge planet to be there because you knew giant planets could not form that close to their stars. But the signal kept repeating regularly, every four and a half days.

Shadows in the Dark: Gas Planets and Cotton Candy

Our solar system was all we knew when we went looking for exoplanets, so we assumed our planetary system was typical and that this was what systems everywhere would be like. Maybe a bit bigger or smaller, maybe with slightly more or fewer planets, but generally, we expected copies of our home system. That is a good starting point if it is all you know. However, that was not at all what the cosmos had in store.

The discovery of 51 Pegasi b was exciting, but it had one glaring problem: it made no sense. Nothing in our solar system even suggested that giant gas planets could survive so close to a star. A planet that circles its star in less than a week would have to whiz around the star at the incredible speed of nearly 300,000 miles (~ 500,000 km) per hour. That is about ten times faster than Jupiter. This planet beats the record of the fastest fighter jet on Earth by a lot—it is 70 times faster. Earth moves around the Sun at about 65,000 miles (~ 100,000 km) per hour, a snail's pace compared to 51 Pegasi b.

Was the signal really a planet? Maybe this star was behaving in a new, weird way. Or maybe it only looked as if it was wobbling. Many guesses created a lively discussion. Was the signal real? And if it was, was it the only one in the cosmos? Was the instrument even working properly?

The discovery of 51 Pegasi b opened the floodgate to new questions. Teams around the world pointed their telescopes at that and other stars, and they found more and more of these wobbling stars pulled around by hot planets that circled them in days. But the question of why these stars wobbled still remained, because astronomers believed there was no way for these gas-giant planets to form that close to stars. Did that mean that 51 Pegasi b was a gigantic rocky world? But where could you get enough rocky material to make a giant rock so close to the star? When astronomers looked at other stars and the disks around them where fledgling planets were forming, they didn't see enough rocks to build a giant rocky planet. So 51 Pegasi b had to be a gas planet like Jupiter. But in that case, how did it get so close?

Although 51 Pegasi b's signal repeated like clockwork, changing our worldview took an avalanche of evidence. What the discovery of 51 Pegasi b—and of many hundreds of exoplanets since—showed us was that the configuration of our solar system is only one possibility among thousands. Our worldview wobbled because we had assumed our solar system was the norm. Our Jupiter feels a nice, steady pull from the Sun but nothing that would inspire it to give up its place and whiz ever closer to its host star— unlike 51 Pegasi b. It took the discovery of dozens of exoplanets before most scientists were convinced that these signals were real and that they were missing a critical part of the story of how planetary systems are made.

What part of how worlds formed had scientists missed? If hot Jupiters can't form where they are found, then they must have moved after. Apparently, some planets wander. How could that work? The star's gravitational pull does not change, so if a planet starts to wander, it should either crash into the star or be flung into the darkness of space. But 51 Pegasi b showed that not all who wander are lost. Some planets stop their wandering close to the star but not so close that they fall into it; instead, they create a stable path and circle the star in quiet continuity until the star expands and changes everything again. But more on that later.

If these planets were real and had formed farther away from their stars but ended up closer to them, then planets must be able to move after they've formed. A planet is born in a disk of colliding material. The gravity of the disk can tug on a newly formed world and change how fast it moves. But the disk exists for only a few hundred thousand years, which is not that long compared to the billions of years that normal stars shine. In that short time, the disk can send the planet away from where it started, like a piece on a round chessboard.

If a planet slows, its star's gravity pulls it closer. If it speeds up, it can move away from the star. That is how planets can end up somewhere completely different from where they started—they migrate. But giant planets plowing through a nascent planetary system would wreak havoc on other planets. If a massive planet like Jupiter plowed all the way to its star, it would throw smaller planets—like Earth— out of its path. The displaced planets could be catapulted out of the system to become lonely wanderers in the darkness of space or shot right into the star to meet their end in a blaze of glory.

Once the star ignites nuclear fusion in its core—when it starts to fuse hydrogen to helium—a strong stellar wind blows away the disk, and the planets settle down. The exception is if one planet crashes into another, but those collisions happen early on, when planets and rocks find themselves in similar paths around a star. By cataloging thousands of planets circling other stars, we can see snapshots of this evolution. Imagine knowledge like a vast net connecting us to everything around us. This net expands when we find something we did not expect—like 51 Pegasi b—and add it to the fabric of knowledge.

Currently, we know of more than five thousand exoplanets on our cosmic shores. This translates into finding one exoplanet every other day since that first discovery in 1995. And the signals of an additional ten thousand candidates are being vetted by teams all over the world to ensure they are the signals of real planets, not measurement errors. Based on experience, eight out of ten of these candidates turn out to be real planets, so if we include them in our count, that means that we have found at least one new world a day since Mayor and Queloz discovered the intriguing signal of 51 Pegasi b.

In science, progress can make turns and twists. What was impossible yesterday becomes reality today and changes what we can achieve tomorrow. Mayor and Queloz's discovery was such a step for science. They were awarded the Nobel Prize for their discovery. One day, astronomers could only speculate on whether there were other worlds circling distant stars, and the next day, we knew there were. Those more than five thousand confirmed new worlds are just the ones in our cosmic backyard, because the closer a star is, the easier it is to discover whether it has planetary company. These new worlds are the first destinations on our map for future exploration of the cosmos.

The scorching 51 Pegasi b is not even the closest or hottest world discovered, and it's by no means the weirdest. Take another exoplanet, WASP-12 b, which was discovered in 2008 by the SuperWASP planetary transit survey team (WASP stands for Wide Angle Search for Planets, a search done with an array of small robotic telescopes). It circles its star, which is tugging it ever closer, in only a little more than one Earth day. In another three million years, WASP-12 b will crash into the star's surface and be consumed. And the scorching planet K2-137 b, discovered in 2017 by the Kepler team, needs only 4.3 hours to circle its star, so for every day on Earth, K2-137 b completes five orbits around its star—so on K2-137 b five of *its* years have passed during 24 hours on Earth.

Compared to the hottest exoplanets, the hottest planet in our solar system is only lukewarm. The wilting heat of Venus's surface of about 900 °F (~ 480 °C) is cool relative to the 8,000 °F (~ 4,500 °C) in the roasting atmosphere of KELT-9 b, about seven hundred light-years away from Earth. KELT-9 b was discovered in 2016 by the KELT (Kilodegree Extremely Little Telescope) team, and it remains one of the hottest exoplanets ever discovered. These first discoveries might make you believe that huge, sizzling new worlds are all that is out there and that our temperate Earth is really special.

In science, it is critical to know what you *can* and *cannot* find and then compare that to what you've found. When you consider that small planets are much harder to find than large ones, the discoveries to date tell a different story. These scorching new worlds hint at an incredible diversity of exoplanets yet to be revealed, but they show only part of the planet population, the part astronomers *can* find.

Astronomers continue to discover more and more stars that wobble back and forth every few days. What does it take to convince scientists that this wobble is really a planet pulling on its star? It takes a different way of finding the same planet. An *independent* confirmation.

If you use two independent methods and obtain the same conclusion, you have confirmed the result. In 2000, two teams, one led by the Canadian-American astronomer David Charbonneau and one led by his compatriot Gregory W. Henry, found something that changed the debate about whether these planets are real. HD 209458 b is about 160 light-years away from Earth in the constellation Pegasus. Its star wobbles, and its light dims ever so slightly at exactly the time when the wobble measurements show that the planet is in our line of sight, as it *transits* across the face of the star. When you walk into a bar and look for your friends, the easiest way to see despite the glare is to block out part of the bright overhead lights. You need to place your hand exactly between your eyes and the light to be able to block the glare from your point of view. You shade your eyes to be able to see more in the room. When you move your hand away, the overhead lights appear as bright as before.

A planet that moves between us and its star blocks part of the star's light for a few minutes to a few hours. Instead of searching for the wobble of a star, in this method, astronomers look for a change in brightness: Does the star appear just a little less bright now than it did before? And does this dip in brightness repeat like clockwork? Based on how much dimmer the star appears, astronomers can figure out how big the object blocking part of the glowing stellar surface is.

The planet HD 209458 b was among the first dozen

worlds detected around wobbling stars. And because these hot Jupiters are so close to their stars, the chances are high—about one in ten—that they will block part of their host stars from our view on their path around them. Not every planet blocks starlight from our view because we need it to be in a line of sight with Earth. So for every ten stars that host a hot giant, nine will shine unchanged, but one will dim periodically, revealing its companion. Seen from Earth, that tenth star's planet casts a dark silhouette against its bright star the way your hand does when you make shadow puppets on the wall.

Every three and a half days (eighty-five hours), like clockwork, HD 209458 b's star appears to dim by a tiny 2 percent for three hours. HD 209458 b blocks part of the star's hot surface about one hundred times every Earth year. And because it is such a small change, the object blocking the light must be much smaller than the star. Small as in the size of a planet. And when you see its star wobbling too, you have another piece of the intriguing puzzle of what this new world is like. The wobble reveals how massive the planet is and the transit shows its size. While we still can't actually see the planet itself, this information gives us its shape and heft. Is it like Earth, dense like a rock, or like Jupiter, a fluffy gas ball? Combining the information from the two search methods shows that HD 209458 b is a hot, fluffy gas ball like Jupiter. If you dropped HD 209458 b into a huge cosmic bathtub, it would float. It has only about half the density of Saturn. Its density is actually close to a marshmallow's—a roasted marshmallow, in this case. But the planet is not made of marshmallows; it is made of mostly hydrogen. And the scorching heat of its nearby star bloats it even further. This hot Jupiter is stranger than fiction—but it answers the question of whether exoplanets are real.

HD 209458 b also revealed what the kind of planet 51 Pegasi b was. If HD 209458 b was a heated gas ball, 51 Pegasi b could be one too. And those two worlds were only the first of hundreds that have confirmed this: hot gas balls exist close to their stars.

You might wonder why astronomers are not a tad more creative when choosing names. HD 209458 does not conjure up a fantastic world with giant storms scorching it. The naming is purely practical—*b* means that it is the first planet found around the star HD 209458 (the letter *A* is reserved for stellar companions in multiple star systems). Letters—*b, c, d, e, f*—make it easier to figure out how many planets are in a system and which ones are the hotter ones, close to the star, because those are generally found first. While the star's name looks like a somehow random combination of letters and numbers, it tells a story. Take the star HD 209458: it got its name from a 1920s survey of more than 220,000 nearby stars, the Henry Draper Catalogue (HD). The name indicates that the star is bright—it is the 209,458[th] entry in the HD survey—and it tells astronomers where they can find more information. In addition to the name HD 209458 b, it also got the nickname Osiris, after the Egyptian god of the underworld. Not all exoplanets have nicknames yet, especially not of ancient gods—with over five thousand exoplanets, you're going to run out of deity names for their host stars pretty fast.

However, *you* can help and propose names for exoplanets via the NameExoWorlds campaign by the International Astronomical Society that I had the honor to kick off in Honolulu in 2015 at its General Assembly. See the chapter "To Learn More" on page 265 if you want to learn how you can name exoplanets.

What would you name one of these newly discovered worlds on our cosmic shore?

Snapshots of New Worlds

Astronomers found more and more exoplanets, but for a long time, there were no photos of those new worlds, only artistic impressions of how they might look. But in 2008, the Canadian astronomer Christian Marois took a photo of a young family of planets circling a star called HR 8799. This image of four small dots of light is the first snapshot of an exoplanet family. HR 8799's practical name comes from being the $8,799^{th}$ star of the Harvard Revised (HR) Photometry Catalog. The star—about one and a half times the mass of the Sun and about five times as bright—is around thirty million years old and about 130 light-years away from us, at the western edge of the Great Square in the constellation Pegasus, the winged horse.

Young planets are hot, still smoldering from the collisions of pieces that formed them. Astronomers can spot them because they are still hot enough to be viewed over the glare of their stars. As these planets get older, they become cooler and fainter, and eventually they disappear from our sight, making the picture of the four young exoplanets a precious memento, like a baby photo you'll keep forever.

Finding new worlds requires looking at the right cosmic moment in order to spot them. These discoveries are unique snapshots in time in an ever-evolving cosmos where stars and their planets move, live and die.

Too Hot to Handle

You run. You push yourself to go just a bit faster, even though you are exhausted. Behind you, the sun is rising, and that sunlight brings with it a heat that will incinerate your bones and super-heat the air you need, stealing your last breath. You run. You try to outrun the sunrise in the little sliver of dusk that races around the planet, bounded on one side by the boiling, superheated day and on the other by the tendrils of the frozen night air. You run. You run in the narrow temperature band that allows you to survive . . . forever outrunning the sunrise.

Could this happen? The idea lies at the heart of the 2004 movie *The Chronicles of Riddick* directed by David Twohy; the sci-fi character envisioned Riddick outrunning the deadly sunrise after escaping from the underground prison on the fictive prison planet Crematoria. Twohy loves science, and the scenarios on this exoplanet are well thought out. Riddick outrunning the sunrise on Crematoria leaves a powerful impression. And when I cheekily wondered what exactly Riddick would be breathing while running, David gracefully admitted that the whole movie would have looked much less interesting if Riddick wore a gas mask the whole time. Pro tip for the cosmos traveler: when rocks evaporate around you, try not to breathe in.

Watching Riddick run through a hellish landscape immerses us in an alien world and gives us just a glimpse of fascinatingly bizarre environments that could exist somewhere out there. For now, imagining these brand-new worlds is the closest we can come to visiting. But astronomers have already spotted rocky planets that are so hot, they could serve as a real version of Crematoria.

CoRoT (Convection, Rotation et Transit Planétaires), a small French-European space mission with an eleven-inch (27-cm) telescope on board, found the first blazing-hot rocky exoplanet in 2009: CoRoT-7 b, another shocking surprise. Its star, CoRoT-7, is just a bit brighter than our own yellow sun and is located about five hundred light-years away from us in the constellation Unicorn (Monoceros). Unlike on Earth, where the sun looks like a small disk to us, on CoRoT-7 b, CoRoT-7 looms large in the sky. CoRoT-7 b is a rocky planet a bit bigger than Earth, and it's sometimes called Earth-like because it is a rock and it is Earth-sized. But CoRoT-7 b is completely different from Earth. Its surface is scorching hot, an estimated 3,500 °F (~ 2,000 °C). It is so hot that rocks melt, evaporate, then rain down again on this lava world. That is somewhat like the water cycle on balmy Earth. But waterdrops or snowflakes hitting you is one thing; rocky drops pelting you is another level of natural hazard. In a storm on CoRoT-7 b, you would need more than an umbrella. An advisory, perhaps in *The Hitchhiker's Guide to the Galaxy*, would read: *"Excellent lava-kite-surfing but don't get hit by the rocky rain."*

But what about the region transitioning between the day side and night side? Could there be a balmy temperate zone where you could outrun the sunrise? Let's see. First, the air would be filled with toxic fumes, so don't breathe. An oxygen mask is a must for prospective tourists. Then the question is "How thick is the atmosphere?" On Earth, day-side and night-side temperatures are about the same. It is a bit colder at night, but not by much. The air and the oceans on Earth transmit the heat around the globe, so whether the Sun is illuminating the sky or not, the temperature is pretty much the same. On other worlds that have winds and oceans, day and night temperatures should be similar too. And while the oceans on CoRoT-7 b are made of lava, not water, they should also transmit heat around the globe, along with heavy storms. When astronomers first found CoRoT-7 b, the possibility of a warm sliver there—like the region Riddick runs in on Crematoria—was central to the discussion of whether this first rocky world could host life. Could such a sliver even exist on such a hot world? I pointed out, to the dismay of some of my colleagues, that if you wanted air to breathe, that would lead to winds transporting the blazing heat everywhere. Another lesser but nevertheless intriguing problem would be that organisms would have to start running immediately and never stop to stay in that warm sliver before the planet is tidally locked in synchronous rotation. But even then, winds and lava oceans should keep both sides scorching hot.

Even if it doesn't have life, CoRoT-7 b would be included on a tour of amazing planets. There is a second small red sun in CoRoT-7 b's sky, gifting you two shadows on this strange world. Watching vast oceans of lava roll under the light of two suns would make this a very worthwhile tour!

Let's leave the idea of Crematoria and lava worlds behind for a minute. Would you be able to outrun the sunrise on Earth? I ask my students to figure this out in my introductory astronomy class. (It might come in handy on an interstellar journey one day, and it teaches them that there is more than one solution to a tricky problem.) Let's take the fastest human as our example. Usain Bolt can run about 27 miles (45 km) per hour—admittedly only for a short sprint. Is that enough? Let's check. The Earth has a radius of about 4,000 miles (~ 6,500 km). One trip around Earth's equator is about 25,000 miles (~ 40,000 km), and you have twenty-four hours to run it. So you'd need to run about 1,000 miles (1,600 km) per hour to make it. That is faster than a normal airplane and just a tad slower than the fastest fighter jet. But actually, you can outrun the sunrise on Earth. You don't even need to be a superathlete to do it. It all depends on where you are. If you start running on Earth's equator, you'll need a jet to do it. But a planet is a sphere, so the closer to the poles you get, the shorter the distance you need to run. Close to a pole, you can take a leisurely stroll away from the sunrise. (The Earth's axis is tilted by ~ 23.5 degrees, so this hypothetical does not work perfectly because in reality, our poles are misaligned from where the sunlight hits. But for the sake of simplicity, let's ignore the Earth's tilt.) This problem shows my students that it is important to think about how you can beat the odds, especially if the odds are strongly stacked against you.

Let's get back to real exoplanets. Are lava worlds made of rocks, like Earth? Or are the building blocks of these worlds different? To check, scientists need to study the rocks. But how do you get a sample from a lava planet when you can't travel there to collect it? Well, I decided that if we couldn't get there, I'd make my own. But how do you make your own worlds if you are not in Douglas Adams's sci-fi classic *The Hitchhiker's Guide to the Galaxy?* (The entertaining story chronicles the misadventures of the only human who survived the Earth's destruction and includes a visit to a planet-making factory.)

At Cornell, I often start my introductory class with a fun cartoon: "The Difference" from the *xkcd* comic strip. It shows a stick figure pulling a lever and getting zapped. Then the cartoon splits in two. On one side, a figure labeled "Normal Person" says, "I guess I shouldn't do that." On the other side, a figure labeled "Scientist" says, "I wonder if that happens every time." I love the cartoon because it makes me wonder why anyone would not want to know if the zapping happened every time (it does not seem to hurt, so why not try it again?). But surprisingly (to me), many of my nonscientist friends don't feel that way. Making my own lava world just required finding the right lever to pull to make it happen.

After long discussions, one of my colleagues and I figured out how to make lava worlds. You start by mixing the right chemicals to create different kinds of rocks, then you melt those rocks. This creates hot lava that could cover the surface of a world far away. I admit, it's not the same as going and collecting a sample, but an artificial lava world is as close as we'll get for now. To do this, my colleague the Costa Rican volcanologist Esteban Gazel and I set up a Lava-World Lab at Cornell in a collaboration between the geology and astronomy departments.

When I walk into the newly minted lab, no rivers of lava greet me (luckily). It's just a big room with lasers, furnaces, and a large brown box that houses our spectrometer. That an instrument that can measure the variation of the light characteristic of a sample, it can split white light and measure its individual bands of color. In addition, the room holds small and large microscopes for measuring the properties of rocks—and whole new worlds.

This is where we make new worlds. The worlds we create are so small, they can easily fit in the palm of my hand. I always wanted to hold a whole world, and now I can hold a couple of them. To melt these tiny worlds, you don't need to generate rivers of lava (that would be quite dangerous). You need only a bit of powdered rock mix and a strip of heated metal to turn these powdered rocks into tiny lava strips. To be precise, the heated strip melts the powder, which forms a "glass." Now, when geologists talk about *glass*, they mean cooled magma. When astronomers talk about glass, we mean the see-through material that comprises the vessels we pour liquid into. This initially led to some confusion at our interdisciplinary research team meetings. Astronomers couldn't understand why the geologists kept insisting there would be glass on the surface of our cooling lava worlds. Finally, when I called it a "Cinderella glass-slipper planet," the geologists understood what the astronomers were thinking. A lot of teasing ensued, but we had arrived at a common language.

Every field has its own vocabulary, and the same word can mean different things to scientists in different disciplines, not to mention to nonscientists. For instance, when astronomers talk about metals, they mean any element heavier than helium. Does this mean that aliens are made out of metal? According to the astronomers' definition, humans are made out of hydrogen and metal. So astronomers better not throw any rocks at the proverbial glass house. You need to know the vocabulary of a scientific field to communicate effectively. Colleagues in your field learned the same language that you did, so they get it. It's like referring to an often-told story among a group of old friends. Everyone knows what you mean, but if there are new people around—like colleagues from another area of science— they'll lose the thread of the conversation fast unless they're brave enough to ask what the heck you're talking about.

The point is that it's important to ask simple, maybe even stupid-sounding questions early on. You might think when you say glass and metal, other scientists know what you are talking about, but they might not. (The same principle applies to life in general, I've found.) I learned to be brave enough—and smart enough—to ask simple questions of scientists in other fields from an expert: Jack Szostak, a Canadian-American biologist and Nobel Prize winner in physiology or medicine in 2009, who was then part of the Origins of Life team at Harvard. Jack was always the first one to ask a question after a talk, and he made it a simple one so everyone else would feel comfortable asking questions they thought they should already know the answers to but didn't. When senior people ask basic questions, junior scientists feel like they can do the same without worrying about sounding stupid. After all, if a Nobel laureate can ask about something so simple, why can't you? I think asking those fundamental questions is part of why Jack got the Nobel Prize. His questions come from wanting to really grasp the basics, and then he uses those basics to understand the world.

Making your own planet is a bit like an extremely difficult chemistry experiment. We chose twenty kinds of rocks that could make up rocky exoplanets, exploring a wide range of possible compositions. We took different chemicals in powder form and mixed them to get the right chemical composition for the rock (and planet) we wanted to create. Rocks come in many forms and colors on Earth, so I was so excited to see the prototypes—the rocky mix—for the worlds we were creating in the lab. I imagined test tubes filled with stunning colors, from ruby red to ebony black. What I found were test tubes filled with white powder (luckily, they were all labeled). It turns out that if you don't mix in iron, all rock powders look white. Still excited—although a bit wiser in the way of rocks—I had startling visions of white worlds everywhere. I still wonder if iron-less white wonderlands are somewhere out there in the cosmos.

Once we added iron and melted the now different-colored rocks, the tricky part started. How do you collect the light from these lava strips so you can figure out what lava planets would look like to our telescopes? This sounds straightforward, but it is excruciatingly difficult, not least because no one has invented an instrument that can do this. No one knew we would need an instrument to measure the light coming from a tiny lava world. We had to find creative ways to use instruments designed for different tasks. In science, doing something no one has ever done before mostly takes a lot of trials and frustrating errors. That happy "Eureka!" moment, when a scientist makes a breakthrough, is built on hundreds of hours of tearing their hair out because what they were sure would work really didn't.

For instance, when the instrument setup that should measure the heat emitted by the tiny lava worlds keeps giving the same answer independent of what you actually measure. And when, at thousands of degrees, the microscope optics could melt because they were never intended to be this hot. (I realize all that now, but somehow, we forgot to account for it in our complex plans to make this work.) That is where creativity and tenacity come in. Tenacity is as important as creativity when you're on the edge of knowledge—you try something, fail, try something else, fail better, then try again and fail again. You learn, painfully, from your failures what not to do and what might work. Every evening when you leave the office, you take a deep breath and give up—for the moment. Then you start all over the next morning when you open the door because you can't let go of the question. You do this again and again, until either you have exhausted all possibilities or it works. One step forward, one step back, but then one day, what you've learned from all the disappointing steps backward crystallizes into a new path and you take two steps forward.

Back in the lab, we catch the glow of our tiny worlds with instruments that were never meant to measure lava worlds. We are paving the way for astronomers to explore lava worlds vast distances away using only light. Our tiny lava worlds are as similar to those hot, rocky planets in the cosmos as we can make them. We found that different kinds of lava look different to our telescopes, so that means we can analyze the molten surface without ever setting foot on a lava world. We use their light and human ingenuity to make the connection to the lab back here in Ithaca, New York. The hundreds of hours we spent patiently making, melting, and observing these tiny new worlds gave us the key. Now we can catch the light from real lava worlds, compare them to the light from our different lava strips in the lab, and see if there's a match. If we find one, we can identify what the surface of that world is made of. Then we can hold a tiny version of that lava world in our hands—right here in our lab.

To search for life in the cosmos, we'll leave the beautiful lava-world spectacle behind us. Lava worlds should be admired from a safe distance. But there are even more intriguing worlds on our cosmic shore that invite us to explore.

Chapter 6

No Place Like Home

The limits of the possible can only be defined by going beyond them into the impossible.

—Arthur C. Clarke

No Land in Sight

Yes, you can get bad coffee in Vienna. Vienna is known for its beautiful cafés, where philosophers, poets, and scientists have found inspiration over endless cups of fantastic coffee for hundreds of years—specifically, since 1683, when the Ottoman forces tried to capture the city. The invaders were chased off, but they left bags of coffee beans behind. Their invasion of the city failed, but the triumph of coffee was just getting started. The popularity of coffee in Vienna began with a spy—a spy at the imperial court opened the first coffeehouse in the city. If you go to a café today, you'll see that he had an excellent strategy. Listening to people's lively and often private conversations in coffeehouses must have made spying a lot easier—and more delicious.

Coffee is still at the heart of Viennese culture; the café is an extension of your living room, a place to meet friends or just read. And the coffee is always accompanied by delectable cakes and pastries.

So here I was at a conference center in Vienna, having flown in from Boston the night before for a meeting of the European Geological Union. I am not a geologist, but I was invited to give a talk on the link between exoplanets and our own planet, a critical connection I'd pioneered.

That morning, about eleven thousand scientists were trying to find a cup of coffee during the twenty-minute coffee break. The coffee in the conference center was free for attendees, but as I looked at the brownish-grayish liquid in the white plastic cup in my hand, I wondered how I could have been so sleep-deprived that I'd forgotten to pick up a cup of coffee on my way.

As I stood in the vast hall filled with poster displays, pondering the unfairness of life, I heard steps echoing along the corridor. I was pretty much alone because by the time I got through the incredibly long coffee line, the next conference session had already started. When you enter a room late, it feels like everyone speculates about the reason for your late arrival, so I'd decided to look at the posters of new scientific work instead.

Posters at a science conference solve the problem of how eleven thousand people can present their work in one week. If everyone gave a ten-minute talk, assuming eight hours per day, the conference would take about a year. (On the bright side, in that case, scientists could just stay here for next year's conference. But they would not get any other work done in between.) So only a few people give talks, and the majority of the conference-goers show their work on posters, an endless sea of thousands of posters that attendees try to read during the breaks. To glance at a few tens of posters, let alone hundreds, you must have the ability to snake your way through crowds and slip through tiny openings between people. If you can't perfect this art, give up early and focus on one region of posters, then identify the colleagues who scouted different parts of the room and get them to have coffee with you and give you the highlights of what they saw. Poster rooms are a great way to kick-start international, interdisciplinary collaboration—those discussions over coffee lead to many breakthroughs. Different points of view combine to create new ideas and new paths to knowledge. An additional plus is that the person describing the poster information for you is doing it in the context of their knowledge—and you get all this in exchange for a cup of free (if mediocre) conference coffee.

The echoing footsteps indicated that another person had decided to make a foray into the empty poster hall. And it was someone I knew.

William Borucki, an American astronomer working at the NASA Ames Research Center, is a giant in my field, who managed, against all odds, to get the NASA Kepler mission launched. By *against all odds*, I mean that he proposed the Kepler mission and got rejected by NASA four separate times. But Borucki just did not give up. He knew he had an important mission that could find out how many planets circle other stars. It was designed to look at the same region of sky, searching more than one hundred fifty thousand stars at the same time for the tiny brightness changes that unveiled their planets. (And it would find thousands of new worlds.) Borucki kept a small team of a dozen scientists motivated, proposing again and again until finally, after thousands of written pages and ever more sophisticated experiments to show that the technology would work, the fourth time was the charm. Borucki's example shows that part of successful science is having the tenacity to keep going against all odds. The astonishing Kepler mission with its fifty-five-inch (1.4 meters) mirror found thousands of worlds circling other stars and rewrote our understanding of planets.

Borucki is also one of the nicest people you could ever meet. We had met before at a small astronomy conference where I presented my work on how to figure out if a planet could be habitable, but since the Kepler mission launched, in 2009, he had constantly surrounded by people, all curious for the latest news from Kepler, so I did not expect him to remember me. Surprisingly, when he saw me, he smiled and headed over. I remember thinking that maybe he wanted to know where I'd gotten the hard-earned coffee I was holding. I debated whether I could recommend it.

That cold day in Vienna with terrible coffee turned into one of the most exciting days of my life. Borucki told me he'd planned to find me at my talk the next day. During our serendipitous encounter, he shared an intriguing—and well-kept—secret that I really, really wanted to shout from the rooftops of this beautiful imperial city: the Kepler mission had found a new world that was just in the right spot. Actually, not just one new world. The mission had found two small rocky exoplanets in the Goldilocks zone of the star Kepler-62. Borucki asked if I could look at the data and tell him what I thought—could these planets be habitable? When I dedicated my career to the question of how to identify habitable worlds modeling their light fingerprints, no one knew when the first ones would be detected and if it would happen in my lifetime. Just like that, in a cold poster room in Vienna, I became part of the discovery of two of the most exciting exoplanets, Kepler-62 e and Kepler-62 f. I felt like the world stood still for a beautiful moment, and then a new worldview snapped into place, containing real worlds that just might be like ours.

The Deepest Oceans in the Cosmos

Today we know that there are most likely billions of rocky planets circling their stars at just the right distance for life, not too hot and not too cold. But before Kepler-62, although astronomers had found planets in Goldilocks zones, they'd done it using the wobble technique, which gave them an idea of the mass of the planet but did not let them distinguish between rocky planets like Earth and uninhabitable small gas balls like Neptune. Scientists believed that warm, rocky planets like Earth existed, but it was by no means a certainty.

However, astronomers involved in the Kepler mission discovered the two worlds circling Kepler-62 via a different technique, the transit method. As mentioned earlier, when a planet travels across our line of sight, it alters the amount of the hot stellar surface we can see. So by observing decreases in the star's light, we can determine a planet's size. Any planet for which we know both the mass and the radius is a rocky world if it is smaller than about two Earth radii. Kepler-62 e and Kepler-62 f were such small planets. Two rocky, temperate worlds circling another star, news that everyone hoping to find life on other worlds had been sitting on the edges of their chairs waiting to hear. And suddenly, my research to find life in the cosmos went from visionary to practical, from far-fetched to applied, from future-oriented to needed-right-now. That is why Borucki wanted to talk to me—because I had worked on the fascinating question of how to find life on exoplanets before we had any planets to search.

Are we alone in the universe? And if not, how do we find other life-forms? To me, these are two of the most intriguing questions in science. But when I began exploring them, no one knew if other habitable planets existed, and several senior scientists strongly suggested that I should change my misguided research topic. In fact, they told me this more than once—maybe they thought I was hard of hearing. They kept asking why I was working on something I might never find. Scientists throughout history must have always been asked questions like this. Over the years, I got good at offering the skeptics a strained smile and saying nothing.

I had prepared for finding rocky worlds like ours by developing a computer model, a bit like the climate models that predict the weather, to figure out how life might change a planet's atmosphere. How would signs of life on worlds under different suns look to our telescopes? My models are complex mathematical constructs based on Earth's data and history; they give insights into the evolution of rocky planets like ours and extrapolate it to rocky worlds circling other stars. This was why I had been invited to give one of the few talks at a meeting of eleven thousand geologists. Based on all available scientific information, I had bet that rocky planets in the habitable zone existed, but that was what it was, a bet based on an educated guess. (And now, just like that, I'd won this bet.)

Now we had all done it. We had found the first potential Earths out there. Borucki, who would not give up; the scientists and engineers designing and building the Kepler mission; the public that supports scientific quests; and every dreamer who had ever looked up at the stars and wondered—it had taken all of us to get here, to find the first new worlds that could be just like ours.

Later that day, savoring the rich taste of coffee with perfectly steamed milk, my *kleines Schalerl Gold* (small cup of gold), at one of the oldest cafés in Vienna, I imagined what these worlds might be like. Using the café's shaky internet, I connected with my computer back at Harvard and started the model runs to figure out if these planets—if they really existed—could provide warm enough surface temperatures for liquid water and to determine how we could explore these two planets with our telescopes. At this moment, in my mind, I saw two worlds covered in endless oceans and waves that never broke onto a shore. Or maybe there'd be some tiny islands here and there. Would the wind carry the smell of salt from the oceans as it does on Earth? Would there be someone or something feeling that wind on their skin?

The first results arrived early the next morning, and after double- and triple-checking them, I emailed Borucki the exciting news: the two new worlds were just right; two beacons of light on our path to finding other Earths in the cosmos. But—and there is always a *but* in science—the discovery still needed to be scrutinized. It could turn out that the planets we'd thought we'd found were not really there, that errors in the measurements or some mechanical issue had altered the data.

That is how we do science: we find something, then we challenge every aspect of it to ensure we're not seeing something we want to see rather than what is actually there. Every scientist knows the drill. This scientific method lets you distinguish what is there and what is not, step by occasionally painful step. Often you disprove theories you have worked on for years. Theories that would have made the front pages of newspapers worldwide—*if* they had been proven.

This ongoing vetting is why Borucki did not tell the world about the discovery yet; we needed to be sure that the planets were real. We ran test after test to see if the data held up. From that shiny day on, I knew every *ping* of my email could signal doom. That was exactly how it felt. I wanted to look at the message and I also really did not want to. I knew that if it turned out that these signals had resulted from an error in an instrument, I wouldn't actually be losing these planets; they would never have existed in the first place. But I had already formed a strong connection to the first two worlds that could be like ours. And *ping* by *ping*, dreaded email by dreaded email, my planets survived—at just the right distance to keep them cozy.

Kepler-62 is a bit cooler and smaller than the Sun. You can find it in the constellation Lyra, about 1,200 light-years from Earth. Kepler-62 e is the fourth planet from the star and circles it every 122 days. It is roughly 60 percent larger than Earth. The outermost planet of the five circling Kepler-62, Kepler-62 f, circles its star in 267 days. It is roughly 40 percent larger than Earth. Such planets are called super-Earths.

You'd have more birthdays to celebrate on these planets than you do on Earth because they are closer to their cooler star and need less time to circle it. But the surface temperatures of Kepler-62 e and Kepler-62 f could be pretty similar to that on Earth. Whether you could breathe there is completely unknown, but temperatures could be warm and cozy. Twelve hundred light-years away, those planets are circling their star, undisturbed by the excited scientists on Earth who just got their first glimpse of rocky worlds that could be just right for life.

What might such super-Earths be like? We don't have one of those in our solar system; Earth is the largest rocky planet here. Super-Earths could keep more of their water because of their increased mass and increased gravitational pull, so the entire surfaces of these planets might be covered with deep oceans. They could be some of the best surfing spots in the cosmos. Sci-fi films have introduced visions of ocean worlds; Kevin Costner's famously expensive 1995 movie *Waterworld*, for instance, depicts a future Earth where all ice caps have melted, submerging the landmasses. While movies provide beautiful visuals of ocean worlds, in reality, deep oceans would create much weirder worlds. The deeper you dive in an ocean, the more pressure builds around you. When oceans are deep enough, at a certain point, the pressure becomes so high that the water turns solid. The bottoms of these oceans would be ice. Not the cold ice we see in winter on a freezing day, floating on top of the water, but much denser warm ice created by the immense pressure of the ocean above it.

Imagining life on other planets is pure speculation for now. But there is no compelling reason why life could not exist on ocean worlds, whether liquid water covers the surface or thick ice sheets like on Enceladus and Europa hide it. Maybe, instead of starting in shallow ponds on a rocky surface—the current popular theory for the origin of life on Earth—it could start in a shallow pond on an ice shelf. I vividly remember sitting in my office at Harvard with papers strewn all over the tables, talking a mile a minute with the Bulgarian astronomer Dimitar Sasselov, an amazingly creative colleague and the director of the Harvard Origins of Life Initiative, both of us excitedly trying to figure out what these super-Earth oceans might be like. Positive curiosity brings out the best ideas, and slowly the image of these ocean worlds took shape in our imagination. The deeper you sink into the seemingly endless ocean, the darker it gets; the water swallows up the red starlight above. You sink farther and farther into the unknown. The pressure builds around you until your hand touches solid ice instead of water—the bottom of the sea. Unfortunately, you'd be crushed way before that, and the water in your body would become high-pressure ice, so diving too deep in these oceans is not recommended.

In some sense, life would be more sheltered in deep oceans, because the deep-water layer would protect it from harmful UV radiation. But life might also never leave the oceans. How would life have evolved on Earth if it had never come to land? The octopus-like creatures, heptapods, envisioned in the Ted Chiang's 1998 science-fiction novella *Story of Your Life*, the basis of the 2016 movie *Arrival*, cross my mind when I think about large ocean worlds. In that thought-provoking story, an alien civilization visits Earth, and teams of scientists try to learn their language in order to communicate with them.

In addition to providing an excellent alternative to the little-green-men visual, the story highlights the often simplified and overlooked problem of how we would communicate with an alien civilization, which we talked about earlier in the book. Misunderstandings are widespread even among people speaking the same language. A galaxy that apparently never had to worry too much about languages is far, far away—and fictional; George Lucas's vision of other worlds in *Star Wars*. We'll get back to whether planets from science fiction could be real a bit later.

The Star Next Door and Its Planet

One of my favorite planets is right next door, astronomically speaking. We talked about its star, Proxima Centauri, earlier in the book when we were discussing interstellar travel. About four light-years away from us in the constellation Centaurus, Proxima Centauri is the easiest destination for us to reach once we invent ships that can travel those vast distances. A red star about the same age as our Sun, it is part of a triple-star system, consisting of two yellow suns, Alpha Centauri A and B, and a red sun, Alpha Centauri C (Proxima Centauri).

Proxima Centauri, the closest star to our Sun, has a planet that could be just right: Proxima Centauri b, a planet in the habitable zone circling this red sun next door to us. It was discovered by the Spanish astronomer Guillem Anglada-Escudé in 2016, and it has been one of the most intriguing recent finds.

Proxima Centauri wobbles just the right amount. Its planet takes only eleven days to complete its path around its red, active sun, which bombards it with flares of intense radiation. The short time it takes to circle its star means that the planet is likely tidally locked, captured in synchronous rotation. That means that only one side of the planet ever sees the sun; you would have to walk away from the sunlit parts of the planet to experience dawn or dusk and then trek even farther to reach the part of the planet that is shrouded in perpetual darkness. Proxima Centauri b, our neighboring planet, could be just like the world described on the opening page of this book.

The red sun might host two other planets as well, but neither of those would be an interesting candidate for life. In the movies, it is not the red sun in the Alpha Centauri system but the yellow sun, Alpha Centauri A, that attracts the most interest. Alpha Centauri A and Alpha Centauri B circle each other every eighty years and appear like one star when seen from Earth by the naked eye—the third brightest in our night sky. Proxima Centauri circles the pair roughly every five hundred thousand years. Neither Alpha Centauri A nor Alpha Centauri B has known planets, but both have inspired the imagination of sci-fi writers for decades. If one of the two stars had a planet, that planet would see two suns in its sky (and a third dim red one very far away). In the 2009 3D-movie *Avatar*, a film written and directed by the science enthusiast James Cameron, the lush, inhabited fictional moon Pandora has an atmosphere poisonous to humans. It is a little smaller than Earth and circles a fictional gas giant, Polyphemus, around Alpha Centauri A.

The idea of habitable moons as abodes for life is based on our hope of finding life on some of the moons in our own solar system. And if a habitable moon were massive enough—like the fictional Pandora—it should be able to provide environments similar to Earth's if it gets comparable amounts of starlight. One difference between a planet and a moon as a habitat is that the moon exists in the surroundings its planet creates. Europa, Jupiter's icy moon that carries the hopes for an underwater biosphere in our own solar system, is constantly bombarded by intense radiation because it plows through the radiation belt of electrons and ions trapped in Jupiter's magnetic field. For Enceladus and Titan, around Saturn, radiation is less of a concern because of the planet's weaker magnetic field, but Saturn's beautiful rings suggest that its gravitational pull ripped some nascent moons apart. Still, considering habitable moons significantly increases the number of potential places where life could thrive. We do not know yet if habitable moons exist, and whether any of them could be home to life, but the possibility is intriguing.

Finding moons provides an extra challenge for astronomers; the tiny signal a moon would introduce as a variation on the small signal of the planet is even harder to spot than the planet's signal itself. But astronomers are not giving up looking for exomoons. Just because astronomers have not succeeded yet does not mean there are no habitable moons out there to find.

One Earth, Two Earths, Three Earths, Four Earths?

If I could have any type of planetary system I wanted, I would wish for one with more than one Earth. I wonder how much farther along our space-travel capacity would be by now if there was another habitable world orbiting our Sun, let alone several. A system with more than one Earth-like planet also allows us to test our understanding of how "Earths" work. It would be the perfect laboratory.

In 2017, the Belgian astronomer Michaël Gillon and the TRAPPIST (TRAnsiting Planets and PlanetesImals Small Telescope) team found a fascinating planetary system about forty light-years away from Earth, orbiting an unassuming small red sun that is about seven billion years old. It is one of about sixty red suns that have been regularly monitored for brightness changes with a 24-inch (60 cm) telescope called TRAPPIST-South, located at La Silla Observatory in Chile. The star TRAPPIST-1 hosts seven Earth-sized planets, three of them orbiting in the habitable zone of their star. These seven Earth-sized planets circle their red star at different distances, making them the perfect Goldilocks test cases. This is what we'd expect from these worlds: planet TRAPPIST-1 b (circles its star in 1.5 Earth-days): way too hot; TRAPPIST-1 c (2.4 days): too hot; TRAPPIST-1 d (4 days): quite hot; TRAPPIST-1 e (6.1 days): just right; TRAPPIST-1 f (9.2 days): just right; TRAPPIST-1 g (12.3 days): almost just right but a bit cold; TRAPPIST-1 h (18.8 days): probably too cold.

So, a whole year on these planets is only between 1.5 to to about 19 Earth-days long. Because the planets orbit so close to their red sun, they could be synchronously locked, with one side always facing the red-orange sun and the other one dipped in perpetual darkness. TRAPPIST-1 e sits in the middle of the seven Earth-sized planets, with three planets closer to TRAPPIST-1 and three planets farther away. From the surface of TRAPPIST-1 e, its sun would appear like a reddish-orange disk, about four times the size of our Sun. The three inner planets could block out a small part of the starlight every few days when they pass between TRAPPIST-1 e and its red sun.

This planetary system is so tightly packed that planets TRAPPIST-1 d and TRAPPIST-1 f would both appear as large as our Moon in the night sky of TRAPPIST-1 e at closest proximity, when the planets are at the same side of the red sun TRAPPIST-1 c would appear nearly as big as our Moon, and TRAPPIST-1 b and TRAPPIST-1 g about half as big when they are closest. TRAPPIST-1 h would appear the smallest of the planets in the system, only about a fifth of our Moon's size. Just imagine a sky full of planets the size of our Moon—and those closer to the parent star would show phases too, just as our Moon does. From TRAPPIST-1 e's point of view, the inner planets should create a mesmerizing show of planetary phases as they circle this red sun. And you could see our Sun from the planets circling TRAPPIST-1—as a beautiful yellow star on its night sky.

We don't know yet what these planets are like, but the JWST will find out. It is observing the TRAPPIST-1 system. Right now. While I type this. The data will get beamed down and run through the strings of computer code called a data pipeline to unpack its secrets. It is incredible to me that way above my head right now, the JWST is peering into the atmosphere of these worlds and exploring what they are made of because we told it to. It is human ideas that make this majestic telescope move. It will take a while for us to acquire enough data to determine what the air of these worlds is made of, because even though the TRAP-PIST-1 planets in the habitable zone circle their star on such tight paths, astronomers can't observe them every time they pass between us and their star. Sometimes the Sun gets in the way, so we need to protect the sensitive JWST detectors from the harsh sunlight. And inconceivably—to me at least—the telescope has more to do than explore new

worlds for signs of life; it's also helping scientists understand for example how galaxies form and how black holes
work. Exoplaneteers get only part of its time to observe
new worlds. But we already started to explore the atmosphere of new worlds.

A World on the Edge

Growing up, I heard over and over again that math and
science were hard and boring. I heard this from other students in my school, from TV series, and from conversations between adults. That is the stereotype, but it's wrong.
Often the link between our world and math and science
is missing in these conversations. You could not find your
way using a Global Positioning System (GPS) without
Einstein's relativity theory. Because the GPS satellites circle very high above Earth's surface and they move at high
speed, relativity explains how fast time runs on board the
spacecraft. You could not use a cell phone or computer
without math and science. Cars, planes, electricity, energy,
medical equipment—the list of math- and science-based
things we use every day is astounding. Also, it seems to be
a well-kept secret that science is an international endeavor
and so gives you the opportunity to travel and to learn
from and discuss ideas with interesting people all over the
world. Normally those people do not include rock stars.
So as I'm sitting in the six-thousand-seat concert hall,
watching Brian May of Queen perform his sound check,
the air-conditioning no match for the Armenian heat, I
ponder how I got here.

There are thousands of empty seats—about 5,960 of them—around me. They will be filled tonight, but no one is allowed at sound check except for the musicians and a few scientists who are here for their own sound check later. The Starmus International Festival was founded by the Armenian-Spanish astronomer Garik Israelian and the British musician Brian May, who holds a Ph.D. in astronomy, and the festival celebrates music, exploration, science, and art. It brings musicians, Nobel Prize winners, artists, writers, and scientists together to share their passions with everyone. In 2022 it was held in Yerevan, Armenia.

The night before, I had drinks and pizza on the rooftop of our hotel on the central square with the musicians of Sons of Apollo; we discussed life in the cosmos, the universe, and how scientists figured out that it is ever expanding, then we seamlessly switched from the mysteries of the cosmos to the mysteries of the music that connects us all. I wonder how many more kids would pay attention in math and science class if they knew that even rock stars are fascinated by the cosmos.

I'd escaped to the rooftop because I needed to look at the stars as a break from writing about them. The Belgian astronomer Laetitia Delrez had contacted me because she and her team had found an intriguing world with the SPEC-ULOOS survey (Search for habitable Planets EClipsing ULtracOOl Stars) using telescopes in Chile, Spain, and Mexico: SPECULOOS-2 c. This is the same team that found the TRAPPIST-1 planets we just discussed, a team that is creative and funny in naming their surveys: Speculoos is the name of a traditional Belgian cookie, Trappist is the name of a Belgian beer.

SPECULOOS-2 c (also called LP 890-9 c) circles a small red star about one hundred light-years away from Earth and it is a planet on the edge—on the edge of its star's habitable zone. About seven billion years old, the planet is a bit bigger than Earth and takes a little more than a week to circle its star. When I looked at the data, I realized that this world was either a thriving hot Earth or a desolate Venus. And that it could be either. What made it so special was that it was the missing link between *just right* and *too hot*. Studying it will provide clues to what happens when a rocky world is flooded with more and more sunlight. All stars get increasingly luminous with time, our Sun included. In the far future, about five hundred million years from now, it will get so hot on Earth that the oceans will start to evaporate, leaving Earth a muggy hot world on its way to becoming another Venus. We can do a lot in five hundred million years. One idea to protect our planet, which seems to come straight out of a science-fiction novel, is to deploy big umbrellas that block part of the sunlight. Or humanity could build cities and nature parks on massive space stations that travel through our solar system and beyond, unbound to any planet.

That is where SPECULOOS-2 c comes in. It is a world just on the edge of the zone where oceans should start to evaporate. If it is still a thriving hot Earth, then we have a bit more time before Earth will become truly inhospitable. If it is already another Venus, we might have a bit less time than we thought (still hundreds of millions of years). After the rooftop talk under the stars, I kept working on a model for this astonishing new world.

Here, with the afternoon Sun casting its light over the sports and music complex in Yerevan, listening to Brian May and Ron Thal getting the sound just right, the spectacular sounds of their two guitars is forever entwined in my mind with the images of a new world on the cusp of habitability, caught between Earth and Venus.

Ancient Worlds

In science-fiction movies, young and old worlds cover the cosmic star map that the *Star Wars* and *Star Trek* franchises explore. In reality, how much older than ours can planets be? It turns out, a lot older. That's not surprising because Earth has been around for only a bit more than a third of the universe's lifetime. With every star's death, more heavy material becomes available to make planets, especially rocky ones. The earlier the star formed, the fewer rocky planets astronomers expect. But a very old star system, Kepler-444, near the constellation Lyra, had a surprise in store: not only do three stars—Kepler-444 A, Kepler-444 AB, and Kepler-444 C—orbit each other in this system, but in 2015, the Kepler spacecraft found five hot planets, smaller than Earth, circling Kepler-444 A closely. All five planets complete their journeys around their star in less than ten days, and describing them as *sweltering* is an understatement. They are too hot to sustain water oceans. But these worlds are smaller than Earth, and, as we have seen, planets that are less than twice the size of Earth are rocky. That means that five ancient, rocky planets circle this ancient orange star. The Kepler-444 system is about eleven billion years, more than double our Sun's age. As we saw earlier, that means these rocky worlds were already older than Earth is now when Earth formed.

The Kepler-444 system is not the sole example of ancient worlds we have already spotted. The cosmos hosts many ancient planetary systems. If some could harbor life, they might give us a glimpse into our possible future and tell us what to do and what to avoid. The Kepler-444 system is about 117 light-years away from us. That means its light took 117 years to get to us and in just a few years, the first radio signals from Earth, which started to leak out about a hundred years ago, will reach these hot, ancient worlds. I wonder what life could be like on worlds where it might have been evolving for billions of years longer. We have no reference point for this; our only guide is life on Earth. Whether you think we will be able to find signs of life on ancient worlds depends on if you're an optimist or a pessimist. In my mind images of captivating landscapes envisioned in science fiction rise from the sparse and clinical lines of computer code on my screen where I am running models of Earth's possible future evolution. As it did in the past, the chemical makeup of our air will change in the future unless we manage to put Earth into stasis—in the good sense of the word—maintaining perfect environmental conditions for humankind. We have stewardship of Spaceship *Earth*, so it's up to us to maintain it and stretch the epoch of humanity into the future. Searching for other Earths can provide clues how to do that.

How to Build a Space Mission: TESS

For every space mission you hear about, like Hubble or the JWST, there are hundreds that never made it off the drawing board. Think of it as trying to create one of the best start-ups ever. You need an outstanding idea, one that will revolutionize everything we know, and an excellent team to pull it off, and then you have to convince thousands of people that this idea is the one perfect idea. More than three decades passed between the launch of Hubble and JWST. Smaller telescopes, like smaller start-ups, are slightly easier to launch, but by *slightly easier*, I mean they are launched more frequently than every three decades, but there are a lot more ideas that compete for these slots. The outstanding idea also needs to arrive at just the right time to fill a pressing need. Remember that Borucki proposed Kepler over and over again before it was finally selected, and since then it has revolutionized our view of the cosmos and our place in it. Kepler alone has discovered more than twenty-five hundred exoplanets.

But with all the amazing discoveries the Kepler mission has made, it left us with one glaring problem. To figure out how many planets there were per star, it had to look at hundreds of thousands of stars at the same time. To fit that many stars into the viewing area, they must be far away. Its search area on the sky was about the size of your hand at arm's length. Kepler mission found thousands of amazing worlds on average a thousand light-years away, too far away to explore closely. During my time at Harvard, I brainstormed with colleagues at MIT and all over the country on how to tackle the biggest problem we faced in our search for life: how to find exoplanets that were both habitable and close to Earth. What we needed was a telescope that could search the closest stars in the sky for exoplanets. If you had all the money and all the time in the world, it might be easy to do this. But the mission proposals scientists submit to NASA or any other space agency have an incredibly tight budget, depending on what the agency's overall budget and portfolio look like. We knew that JWST was our first chance to search for signs of life on nearby planets, so we needed to propose a relatively cheap space telescope that could search the whole sky for possibly habitable exoplanets around the closest stars. Basically, we needed to design a space telescope that would create a list of the best targets for JWST to look at. What comes next in any spacecraft design is a delicate balance between ambition and cost.

I joined a tenacious team of scientists and engineers, many of whom had been part of the Kepler mission. We proposed our best plan for a new planet hunter to NASA. We called it TESS (Transiting Exoplanet Survey Satellite). These proposals must follow specific formats, which makes a lot of sense because other scientists read and evaluate them. Working on bringing a new idea to life is exciting, but it has to be fit in somewhere between teaching, research, proposal and grant writing, mentoring, and outreach. With hundreds of ideas competing for each slot, your proposal had better be excellent, easy to understand, convincing, and under the page limit! That is much more constricting than it sounds. I can't remember how many times we wrote and rewrote every individual page, making our case, plugging any holes in our arguments, and answering any questions a reviewer could possibly ask. And that is only the scientific part of the proposal. Then there is the proposed budget, which, luckily, I was not part of at that time. The principal investigators of spacecraft missions have to make sure the science rationale, the work packages, and the people assigned to them are perfect fits and that the budget comes in under the allowable limit. Remember the hundreds of other proposals? Yours will be judged in relation to them, and if yours is found lacking in any item, your chances to get a launch slot will disappear.

We had two years until the proposed JWST launch date, so to search the whole sky for potential habitable worlds in that time allows for about a month per star; the mission scans the whole northern sky in one year, and then the southern sky the next year. Ideally, you want to have much more time per star so you can find planets that take more time to circle it—a planet in the habitable zone around its sun needs an Earth-year; luckily a planet in the habitable zone around a red star needs about a month. But fortunately, the sky looks like a dome to the spacecraft, so stars at the highest point of the dome are in all the photos and you can observe them for a much longer time because the pictures of the cameras overlap there.

If you don't know how many planets to expect per star, scanning the sky is risky. You will miss out on detecting a lot of planets because you cannot see a star's brightness changing only if you are not looking at it. With only one month of observing time, you'll miss any dimming that happens in the other eleven months. But NASA's Kepler had shown that almost every star had a planet. With that many planets in the sky, looking at stars for a short time instead of staring at each of them continuously for several years became an option, freeing us to propose a small but mighty telescope: TESS, the relatively little spacecraft-that-could. It has only four-inch (~ 10 cm) cameras on board, but it searches the whole sky for tiny changes in the brightness of our closest stars, exploring our backyard for new worlds.

The TESS launch was the first rocket launch I ever attended. We made it a family trip; we packed our bags for sunny Florida in the middle of a cold Ithaca spring, leaving our winter coats in the trunk of our car at the airport. With our NASA special-guest QR code for the launch on my phone, we were off. On arrival in Orlando, we made our way to Cape Canaveral and checked into a hotel on the beach. While my then four-year-old daughter and I were building sandcastles, we spotted the launch site, where our little telescope waited for the big day in the distance. The Kennedy Space Center, bustling with astronomers from all over the world, creating a beautiful mix of jet-lagged and semi-jet-lagged colleagues with their kids running around chasing each other. Years of planning, building, and hoping had gone into this moment, and the center was filled to the brim with members of the big team that had made TESS happen. I met many of them for the first time in this beautiful setting framed by sunshine and high expectations. I knew the other scientists, but I didn't know most of the engineers who'd built our telescope. It takes an international village to make a telescope fly. The air was full of anticipation, hope, and a bit of suppressed worry about whether our telescope would blast off successfully.

Our launch was delayed for two days because of a fueling issue, which made us miss our flight back home. But my daughter was not in school yet, and one of my colleagues was kindly managing my class for me that week. So we could stay in Florida to see our little spacecraft off. My students were all excited to get daily updates from the launch (and stickers from the mission).

On April 18, 2018, TESS blasted off from Cape Canaveral on a Falcon 9 to begin scanning our sky for the closest worlds. We sat on the bleachers next to the Apollo/Saturn V Center, which displays the massive Saturn V rocket, connecting our launch to the whole history of space travel. I was on the edge of my seat as I counted down the seconds to launch with my daughter: "Ten, nine, eight, seven, six, five, four, three, two, one." Then the rocket blasted off and quickly became a tiny dot of light in the sky, moving ever faster toward its destination. I vividly remember the hugs and elated smiles from hundreds of people who were part of this moment of hope that bridged time and space and set our mission on its course to create the best list of targets for the search for life close to home: a perfect launch day.

Afterward, my family and I celebrated with good friends in a nearby tiny Japanese hibachi place. The chef set an onion volcano on fire for my four-year-old. If you ask my daughter about our trip to Florida, she'll tell you about the onion volcano. And maybe she'll mention that she also saw a rocket launch Mama's mission to space. Maybe.

TESS is still scanning the sky for the closest planets, including those that resemble our world intriguing travel destinations for future explorers, including some that no one expected.

Planets Around Stellar Corpses:
White Dwarf Planets

In 2020, TESS found a planet that, in theory, should not exist: WD 1586 b, a gas giant circling the husk of a dead star, a white dwarf. That the planet had survived the demise of its star was an extraordinary discovery, and it made us wonder: Could life on a planet survive the death of its star?

Remember, when a star like the Sun dies, the nuclear fusion in its core stutters to a final stop; its core will never again be hot and dense enough to fuse lighter elements to heavier ones.

The energy production of a star is not smooth; the fusion engine stutters and starts again after each step. Think about an old car on ice-cold mornings. It starts and stalls but, after a few tries, eventually gets going—until that day when nothing you do makes any difference. It's dead. For you, in an icy-cold car, that is extremely inconvenient. For a star like our Sun and its planets, it is a disaster. The nuclear fusion in its core stops, and all the heavy mass that always tried to drop down to the core but was kept at bay by the energy output from the fusion now crashes unimpeded onto the core. An atom is made of protons and neutrons at its center, with circling electrons whizzing by. When the mass of the star's outer layers crash onto the core, these electrons get pushed together. But electrons generally cannot be in the same place at the same time, so part of the material that crashes onto the core is bounced out and expelled from the star as the magnificent planetary nebula that contains about half of the mass of the original star.

This leaves behind the exposed scorching stellar core of a dead star like our Sun, which is only a bit bigger than the Earth, a white dwarf. It is all that remains of a bright star that reached the end of its life, whispering tales of a cosmic journey. A teaspoon of a white dwarf weighs about 15 tons. To put that in perspective, an adult blue whale weighs about 100 tons, as mentioned earlier. So, about 7 teaspoons of white-dwarf matter weigh as much as a blue whale.

The dissipation of about half the star's mass disrupts the delicate balance between the star's pull on the planets and the planets' counter-push: planets either crash into the star or get flung out into space. The white-dwarf stage was the end of the story for a massive star's planets—or so we thought, until September 2020.

I had often wondered whether life on a planet could survive the death of its star. Even before TESS found WD 1586 b, my team modeled the conditions that would exist around the exposed core of a dead star, and with it the far future of our own solar system. We found that the Goldilocks zone around stellar corpses can last for billions of years. That could allow life to get started again. But maybe life might even be able to survive the ordeal of its star's death—sheltered somewhere under the planet's surface.

Most stars, like our Sun, will end up as white dwarfs. Imagine a universe in the far future with a vast number of stellar remnants cooling down slowly, dimming in the night sky. Astronomers once thought the future cosmos would be eerily silent because planets couldn't survive the demise of their stars. But maybe that's not the case; maybe the cosmos will teem with newly started life-forms or hardy survivors. That depends on what planets can survive and what life on those planets can survive. How did the gas planet that TESS found get to where it was? The best explanation so far is migration. Most likely, the planet initially circled its star much farther out, survived the death of its star, and then got pulled in. In any event, the planet is now happily circling the white dwarf. It got there somehow; we just have to figure out how. And when we do, we'll be able to add a new chapter to the story of planetary survival skills.

A noisy, life-filled future of the cosmos is back in play, and it might be full of planetary survivors. We don't know yet whether life on these planets can survive, but in September 2020, we learned that planets can survive the death of their star, giving life on them a fighting chance.

Planets Around Even Weirder Stellar Corpses: Pulsars

For stars at least about eight times the Sun's mass, stellar death is extremely violent. Remember that once the nuclear fusion in the core stops and the once-stable balance falters the core collapses inward with unimaginable force. The immense mass crashing onto the star's exposed core is so heavy that it can compress it into a neutron star. The collapsing mass presses the electrons into the atom's center, overriding the force that kept the electrons and protons successfully apart for eons. Electrons and protons are squeezed to tightly packed neutrons. The neutrons finally succeed in pushing back against the infalling mass, stabilizing the high-density stellar corpse—creating a *neutron star* and setting up a spectacular cosmic explosion.

Imagine a rubber ball on the ground struck by a hammer. Initially the hammer compresses the rubber, but eventually the ball's density and pressure stops the movement of the hammer. The hammer is thrown back violently by the recoiling rubber ball. The outer layers of the star falling toward the center correspond to the hammer, the core to the rubber ball, setting the stage for an incredibly powerful collision. This drives an expanding shock wave outward from the neutron star, creating the pressure and temperatures to generate elements heavier than iron. The expanding shell of gas and dust is observed as a supernova remnant like the Crab Nebula, which was first described nearly a thousand years ago as a new star in the constellation of Taurus by Chinese astronomers.

A teaspoon of a neutron star's material weighs about four billion tons. It is so incredibly dense that, a teaspoon of neutron matter weighs about as much as 40 million adult blue whales. When stars are even more massive—about thirty times as massive as the Sun or more—even the neutrons can't win the fight against gravity. Gravity tightens its grip even further on the stellar core and squishes these tightly packed neutrons even closer together, forming a *black hole*.

A black hole is what is left of the core after a supermassive star explodes. It is something new, a singularity in space, with colossal gravity that can far exceed that of even neutron stars. The gravitational pull of a black hole is so enormous that it can capture light and keep it circling its center.

But back to neutron stars. If you look at one just right, a pulse of energy will graze you every time the rapidly rotating magnetic pole sweeps into view. Imagine it like the beam from a lighthouse. This subset of neutron stars are called pulsars. Remember the map on the cover of the Golden Record depicting the position of Earth in respect to pulsars, each with its own signature pulse rate, making them easy to identify, and ideal as reference points on a cosmic map. And pulsars—especially millisecond pulsars—are some of the best clocks in the cosmos.

In 1992, the Polish astronomer Aleksander Wolszczan at Arecibo Observatory in Puerto Rico noticed something odd in the signal of one of those millisecond pulsars 2,300 light-years away in the constellation Virgo. The strange signal came from PSR B1257+12. This pulsar's pulses were just the tiniest fraction off. This was due to three objects, two with about four times Earth's mass and another with only 2 percent of Earth's mass, tugging on the millisecond pulsar. There are lots of open questions about how this weird system came to be and whether those objects are exoplanets, exposed planetary cores, or caught objects that came too close to the pulsar. The explosion that creates a neutron star is enormous, so how these objects came to circle the pulsar is still unclear. In the decades that followed this discovery, only a few more objects around pulsars were found. They seem to be very rare, which does not make it any easier for us to find out how they got there and what they are like.

An even weirder system, PSR B1620–26, was found 12,400 light-years away from us in the globular cluster of Messier 4 in the constellation of Scorpius. It is another pulsar, but it is not alone in its gravitational dance. It has a companion, a white dwarf. The system has an object about twice the mass of Jupiter that circles both stellar corpses. This planet is called PSR B1620–26 (AB) b; the (AB) indicates that it circles two stellar cores, which takes several tens of thousands of years. Hundreds of thousands of generations of humans will have lived and died before a year on PSR B1620–26 (AB) b has passed. Astronomers don't expect these pulsar planets to harbor life because of the massive explosion they survived and the extremely potent radiation of the nearby pulsar. The object PSR B1620–26 (AB) b has another distinction—if it formed with its host, it would be extremely old, about thirteen billion years.

But as weird as these worlds circling stellar corpses are, there are even weirder ones.

Lonely Wanderers

Rogue planets have no stars. They are hurtling alone through space forever, solitary wanderers lost in the vast darkness, without a star to light their way. Such planets probably were kicked out of their star system early on, when collisions between just-formed worlds were ongoing. When the early collision between Earth and a Mars-sized object gifted us our Moon, and luckily, Earth was not expelled from the solar system in the process. Science fiction plays with the concept of fictional rogue worlds like Mongo, where Alex Raymond's 1930s *Flash Gordon* comic strip took place (though how the different thriving climate zones on this fictional planet could be maintained without a star remains unclear). Note that later versions, like the 1996 cartoon series, have the planet orbiting another star.

The first candidate rogue planets were announced by microlensing observations by the OGLE (Optical Gravitational Lensing Experiment) collaboration in 2017. About ten rogue planets have been spotted so far, hinting at a large number of these objects. Some likely have a similar mass to that of Earth, such as OGLE-2016-BLG-1928L b, announced in 2020 by the OGLE collaboration. Not all planets have quiet childhoods that allow them to stay in place as we saw earlier. Many get shoved around by gravity and the push and pull of a young disk, forcing them to migrate. Migration shapes the exoplanets we see around other stars. Some planets migrate inward, some outward, toward the edge of their system and beyond, depending on their speed, and some planets collide. These collisions can get violent, like two race cars crashing into each other. So violent that a planet can be thrown out of its star system and end up forever hurtling through cold—and mostly empty—space.

These ever-cooling worlds wander the universe alone, but are those worlds so cold that they are dead husks of possibilities? Or might life form in the residual heat from the planet's formation, allowing for a temporary window of opportunity? Wandering in the darkness of the cosmos, those worlds will remain a mystery because they don't send out light we can catch to explore them.

Better than *Star Wars*

Imagine you are sitting in a movie theater: The house lights go down, and the screen leads you to a planet in a galaxy far, far away. Vast sand dunes on this fictional planet appear to be a desert. The desert landscape unfolds under the light of two suns, Tatoo I and Tatoo II. This world from the *Star Wars* universe, Tatooine, has fascinated millions since 1977. It is the home of young Jedi Luke Skywalker. That fascinatingly strange world grew out of human imagination, but only three decades after that brilliant image projected on the movie screen, scientists found a real Tatooine-like world. And the idea of planets circling more than one sun is not unique to the *Star Wars* universe; Gallifrey, the fictional homeworld of the mysterious time-traveling Doctor Who, also circles two stars.

About two hundred and fifty light-years from Earth, the real planet Kepler-16 b, discovered in 2011 by NASA's Kepler mission, orbits a pair of stars. Two stars, one orange and a bit smaller than our Sun and one small, red star, create a spectacle in the sky a bit like that in the imagined Tatooine. Unlike Tatooine, though, Kepler-16 b is a real planet like Saturn, a huge gas ball with no solid ground to stand on. We don't know yet if Kepler-16 b has a rocky moon, but if it does, that moon could be a mixture of sci-fi visions—*Star Wars'* Tatooine mixed with *Avatar's* moon Pandora—leading to a more spectacular sky than even science fiction has envisioned so far. (And yes, please go ahead and use this for a new sci-fi story.)

Watching our lonely yellow Sun in the sky, you might find it strange to imagine a second sun there. But about half of all stars come in pairs, so half of all exoplanets should have two suns as well, just like the fictional Tatooine. However, finding planets circling double stars, called binaries, was a surprise because scientists believed the gravitational pull from two stars would have thrown out most of the material that could make planets. But these planets exist, orbiting either one or both of their stars. The combined pull from two stars on a planet changes over time, as does the light it gets from its two suns. There is some speculation by fans of the fantasy *Game of Throne* series that the changing seasons on the continent of the fictional Westeros could be explained by the chaotic tug of two suns in a very specific configuration. But even with two suns, the conditions on Westeros are extremely hard to explain.

For most real exoplanets orbiting two stars, the temperate region is not plagued by chaotic seasons. When two stars tug at a planet, the planet can survive over billions of years only if it finds a stable orbit, either circling one of the two stars with the other one much farther away or circling the pair of stars. In both cases, the light reaching the planet changes only slightly over time, making two suns in the sky a captivating view with mostly harmless side effects. Envisioning the most extreme dance a stellar pair could make and still be bound together, I calculated where the habitable zone region around stellar pairs is, a place where rivers and oceans could glisten on the planet's surface under a

Tatooine-like sky. So yes, Luke Skywalker could have enjoyed a warm day under two suns—but I always wondered where his second shadow was.

Kepler-16 b also demonstrated that we can only see glimpses of the cosmos, a snapshot of the dynamic universe that reveals its secrets in its own time. If we were searching for Kepler-16 b now, we would not be able to find it. In 2011, when Kepler-16 b was discovered, it blocked out a small part of the hot stellar surface of the first star and then the second star. But in 2018 Kepler-16 b moved out of our line of sight; we can't see its shadow anymore! It became, to us, an invisible world, one of the billions in the cosmos we cannot spot. But we know it is there. The dynamic dance of the stars in the cosmos keeps hiding and revealing spectacular new worlds around us. All the exoplanets we have found hint at an incredibly vast diversity of planets out there.

But planets can have more than two suns in their skies. Kepler-64 b is a Neptune-like gas giant about 130 light-years away that circles a double-star system, and that binary orbits a second distant pair of stars, that would shine bright in that night sky. Kepler-64 b was discovered in 2012 by two amateur astronomers volunteering for the Planet Hunters, a citizen scientist program where volunteers from all over the world reviewed astronomical data from NASA's Kepler spacecraft. It was originally designated PH1 b (Planet Hunter 1). You don't have to be a professional astronomer to join the search for new worlds; through citizen science projects, you can help write the map for real missions or future starship *Enterprise* destinations. You can discover your own new world; at the end of the book you can find out how.

Maybe our descendants will travel to these planets in the far future and stand under the light of two suns, watching their double shadows dance.

Science-Fiction Planets to Watch

Most astronomers I know enjoy science fiction, sometimes questioning how to implement its finer plot points. I remember sitting with friends in a movie theater in Cambridge, Massachusetts, watching the movie *Avatar* when it premiered in 2009, and the heated discussion that broke out about whether the fictitious material harvested in the movie, unobtainium, on the fictional moon Pandora could not be better harvested by lassoing one of the floating mountains that contained it and how to do that—until the stares of the nonscientist moviegoers ended our calculations.

An astronomer and good friend of mine, the British-American Jonathan McDowell, lives among bookshelves filled with science-fiction books interspersed with the largest compilation of records of all the spacecraft ever launched—a collection shares online and grows by compiling available records and xeroxing documents in basements of space agencies in all corners of the world. He also owns a sushi utensil set in the shape of the Starship *Enterprise*, and a replica of the blue TARDIS, the fictional time machine of the BBC's Doctor Who. He throws fun, elaborate, Martian-themed birthday celebrations, with an eclectic mix of all red foods—such as red velvet cake, red lentils, raspberries, tomatoes—and with water glasses precariously balanced on Mars bars (because yes, there is water on Mars; it's a bad astronomy joke). Astronomy and science fiction can be a great combination, a quirky, unusual, and entertaining part of life.

Some real exoplanet host stars have a special place in science fiction (and in our hearts). In the highly entertaining 2021 book *Project Hail Mary*, the novelist Andy Weir follows the adventures of a reluctant hero, an astrobiologist turned high school teacher who travels to another planetary system to save Earth. I won't spoil the story here, but the novel includes interesting real host stars with known exoplanets: Tau Ceti and 40 Eridani.

Tau Ceti is a lone star that's about twice as old as our Sun and only a bit smaller; a mere twelve light-years away from us, it's in our cosmic backyard. It is visible to the naked eye in the night sky and located near the celestial equator in the constellation Cetus, the sea monster. From Tau Ceti, our Sun would be visible in the sky in the constellation Boötes, the herdsman—assuming future astronauts will look for the familiar stellar constellation in an alien sky. Because it is the closest Sun-twin, it has been featured heavily in several sci-fi novels, among them Isaac Asimov's *The Caves of Steel* in the 1950s, Larry Niven's *A Gift from Earth* in the 1960s, Kim Stanley Robinson's 2015 *Aurora*, and Weir's *Project Hail Mary*. Four planets have been announced around Tau Ceti, with at least one planet circling it in the habitable zone. While we wait for more information on Tau Ceti's planets, they are already inspiring bold ideas for a future where we can travel beyond our solar system.

The triple-star system 40 Eridani is about sixteen light-years away from Earth and consists of an orange star (40 Eridani A), a small white dwarf (40 Eridani B), and a red sun (40 Eridani C). In the *Star Trek* franchise, the planet Vulcan, the home world of Commander Spock, circles 40 Eridani A. To an observer on a planet orbiting 40 Eridani A, the B-C pair would appear slightly brighter than Venus in its night sky. Our Sun would appear in the constellation Hercules. In Weir's *Project Hail Mary*, a fictional planet called Erid, a hot, fast-orbiting, massive rocky planet, is home to spiderlike aliens, and provides the human main character in the novel an opportunity to make highly entertaining first contact. The fictional Erid is loosely based on a candidate-planet signal around 40 Eridani A.

While there was a first potential exoplanet signal in 2018—indicating a blistering planet could exist circling 40 Eridani A—newer results from 2023 suggest that the signal originates from stellar activity, not from an exoplanet. Astronomers have not yet found any real planets circling in the 40 Eridani system, so as of yet, no real Vulcan or real Erid, which does not mean that they can't exist.

An even closer orange star, Epsilon Eridani, located at about ten light-years away, also features heavily in sci-fi visions of the future; it stars in the 1990s TV series *Babylon 5*, video games like *Halo* and *Race for the Galaxy*, and in novels like Isaac Asimov's *Foundation's Edge* and Alastair Reynold's Revelation Space series. Epsilon Eridani is visible to the naked eye in the constellation Eridanus, the river. From Epsilon Eridani, the Sun would be visible to the naked eye in the constellation Serpens, the serpent.

Epsilon Eridani and its giant exoplanet, Epsilon Eridani b, which is about half the mass of Jupiter and circles the star every seven years, are very young: between one-half and one billion years old. At that age life on Earth was still in its infancy. Epsilon Eridani b is now also called Ran, the Norse goddess of the sea, the name that was chosen from all the entries proposed to the NameExoWorlds campaign.

If there is a yet-to-be-discovered rocky moon circling Epsilon Eridani b or another, still unknown, rocky exoplanet in the habitable zone of Epsilon Eridani, this system can give future space travelers another intriguing nearby destination inspired by visions of these imaginary sci-fi worlds. So far, the exoplanets we have discovered have been weirder and more exciting than we imagined.

New Worlds on Our Cosmic Horizon

With telescopes from the ground and space, we have learned that the universe is teeming with a fascinating variety of planets, more types than we could have imagined. When our ancestors first looked up at the stars, thousands of bright lights caught their attention. Now we know that those bright lights have companions, planets that circle these stars in a universe full of possibilities.

More than five thousand new worlds paint an intriguing picture:

1. Hot Jupiters were the first big surprise—incredibly hot gas exoplanets so close to their stars that their outer layers are partly boiling off. They are easier to find than other planets.
2. Some planets finish a full exoplanet year in less than an Earth day.

3. Most planets are not lonely—most stars have more than one planet (and about half of the stars have stellar companions too). Spectacular double sunsets and sunrises are common on the horizons of worlds circling two suns.

4. Expect rocky rain because rocks should evaporate and rain back down on oceans of magma on the hottest rocky worlds discovered—lava worlds blasted by light and heat from the nearby star.

5. Among the thousands of new planets discovered to date, astronomers already have identified about three dozen rocky ones that get about the same amount of light and heat from their stars as Earth does from our Sun.

6. All of the potential habitable worlds astronomers have found so far see red suns in their sky, because it is generally easier (and faster) to spot exoplanets circling smaller red stars.

7. Some detected ancient worlds were already older than the Earth is now when the Earth formed, about four and a half billion years ago.

8. Several of these ancient worlds have survived the violent explosions of their stars and are now orbiting stellar corpses.

9. Rogue planets are not bound to any star anymore.

10. At least every second star has one planet or more circling it.

11. At least every fifth star has one or more rocky planets circling it in its Goldilocks zone, that prime real estate where liquid water could glisten on a planet's surface.

These are just a few of the fascinating new findings that have shaken scientists' worldview and remade our understanding of what planets can be like. Among the thousands of exoplanets we have discovered, perhaps we have already found the first that could be an alien Earth.

With two hundred billion stars in our galaxy alone, the odds of finding life seem to be ever in our favor.

Chapter 7

At the Edge of
Cosmic Knowledge

Every point in space is the center of its own
sphere of ever deepening time, bounded by
a shell of fire.

—Katie Mack, *The End of Everything*
(Astrophysically Speaking)

The View from Carl Sagan's Office

I see the world around me from a particularly privileged
vantage point, the same view that Carl Sagan, a visionary
astronomer who shared his passion with the world, must
have seen as he wrote his books in this same third-floor
office of Cornell's Space Sciences Building, in a corner that
seems to reach out into the trees and shrubs of this green
hilltop campus, with Cayuga Lake glittering in the distance.
I never met Carl in person. I wish I'd been able to run into
him in the corridor, or in the small coffee room on our floor,

and talk about the new worlds we are discovering. Nevertheless, his work has had a profound impact on me and sparked the curiosity and imagination of everyone at the institute that carries his name, and of so many other people.

The trees outside the window are a bit taller now than they were when he was here; the skirts and haircuts of the students are a bit shorter; and the formulas, ideas, and drawings on the large whiteboard are about exoplanets rather than the Voyager mission's Golden Record. Still, it is largely the same view that Carl had as he observed the university bustling around him and pondered the mysteries of the universe. I suspect that he was more orderly than I am; my bookshelves are overflowing, and slightly unstable piles of research papers cover my desk. But I can feel his curiosity all around me. Standing by the tall window, I wonder if this view of the old oak tree was his favorite too. I like to imagine him coming in through the same office door, standing at these windows, and watching the world around him unfold—a link through time.

It was several years ago, just after I moved to Ithaca, that I founded the interdisciplinary Carl Sagan Institute here at Cornell University, bringing a team of curious minds together to find life in the cosmos. Our first gathering took place in my cluttered conference room. I bribed thinkers from many different departments to attend the meeting by promising them good coffee, dark chocolate, and, most seductive, a discussion of the intriguing question of how we might find life in the cosmos.

From that beginning, our institute team has come to include members from fifteen different departments, representing disciplines as diverse as astronomy, biology, chemistry, engineering, music, science communication, and performing arts. The team members' ideas and opinions are just as diverse, but they are all united in the quest to find life in the universe. Their many different accents and vibrant expressions add color to the lively discussions among eminent scholars, researchers, and students who just started at Cornell and are curious about how scientific research works. Today, two red espresso machines saturate the air with the smell of strong coffee, and the room is filled with warmth and laughter when we get together. These are some of the most interesting afternoons in my professional life.

Science is a rich fabric of knowledge that spans time and place; an invisible net stretching above our heads like a second sky where sparkling ideas stand in for the billions of stars. When I close my eyes, I can imagine the ideas of millions of people connecting us all to those who came before us and those who will come after. Their discoveries, large and small, all aid in our quest to figure out the mysteries of the universe and our place in it.

Asking the right questions is crucial in science because you have only one lifetime to figure things out. But your one lifetime is just one link in a long chain of explorers reaching across centuries. Ideas leave their marks on our world and echo long after the individuals who came up with them are gone. Some of these thinkers' names are familiar—like Albert Einstein and Marie Curie—but others only scientists recall, and uncounted names are lost in time. History is colored by the biases of those who write it, and even narratives of scientific discoveries are subjective. Women and minorities who did not fit the traditional view of what scientists looked like were left out, as is the case in many aspects of history. Times are improving slowly, and some overlooked researchers are starting to be acknowledged. But although many nontraditional scientists' names have been forgotten, their ideas are not lost. They stretch across the millennia and help us uncover the mysteries of the universe.

Some ideas stand the test of time. Others turn out to be wrong or incomplete. People who watched the sky were initially fooled into thinking that the Earth was the center of the universe, until more and more observations showed that they did not have the complete picture of our place in the cosmos yet.

Research is often thought of as stuffy and rigid, but imagination and creativity are the backbone of science. Scientists are adventurers, exploring the frontiers of the unknown, trying to determine how things work in unexplored realms. Imagine finding a dinosaur bone and figuring out from there that it meant colossal creatures once walked the Earth. Imagine discovering that the universe was born in an unimaginably hot and dense Big Bang or that another sun wobbles just the right way every four and a half days to show you the first new world circling another star. Asking questions is what becoming a scientist is about. And some of the answers are written in the night sky.

Cosmic Knowledge

When I look up at night, I see an incredible black tapestry dotted with bright stars. But an understanding of the cosmos makes this picture breathtaking because the black tapestry gains depth and meaning when you grasp some of its mysteries.

The night sky shows us space, but more than that, it shows us the past unfolding before our eyes. Everything you see looking up at night has already happened, but we are only learning about it now, when the information encoded in light arrives. If light did not need time to travel, we would be blind to the past and would never be able to determine the origins of our cosmos.

What you see in the on this black tapestry is "now" only here on Earth. Other places in the cosmos will have to wait thousands of years to see what you see tonight, and yet other places have already seen what we will see in our future. To me, this makes every night special. Because only you and I and our fellow Earthlings get this "now"—only here in our specific place in the cosmos.

The view that our corner of the universe is special is very hard to shake. The idea that Earth was the center of the cosmos held for so long because of what Sagan called the "unlucky coincidence that commonsense observations and what we secretly wished to be true resonated and converged." Looking up at night, we could easily conclude that the stars and the Sun move around us. Transitioning from that comforting view to a worldview where Earth is only one of many planets orbiting one of many stars took time and perseverance. Eventually, humans relinquished the idea that the Earth was the center of the universe and reluctantly admitted that our planet travels around the Sun—but at least *our Sun* was still the center of the universe. Then, as scientists made more and more observations, the idea of our Sun holding a privileged position in the cosmos also collapsed, leaving us on a pretty ordinary planet circling a pretty ordinary star—a worldview that is not as comfortable. It is nice to be special.

But we gained a great deal when we lost our perceived place at the center of the cosmos. Observations of the sky above made over the course of hundreds of years opened our species' eyes to the immense cosmos we are part of and to our place within it. Our home address is planet Earth, the third rock from the star we call the Sun, one of about two hundred billion stars in our galaxy.

We have no photo of the two hundred billion stars in our Milky Way—and we won't have one for a long while. To fit the entire Milky Way into a photo, a spacecraft must fly far away from Earth, and far above the plane of our spiral galaxy. And no spacecraft has even made it to the next star over yet. The Earth is like a piece of pepperoni on a pizza trying to imagine the whole pizza's shape.

One of the main differences between us and the questioning pepperoni is that we have figured out what our galaxy looks like. Astronomers measured the position and the movement of the stars in our galaxy, compared that to photos of thousands of other galaxies in the cosmos, and found one that looks similar. That will be a stand-in until we get a real photo of the Milky Way. My students find the pepperoni pizza idea helpful. Free pizza lunches are common on campus, and now pizza reminds them of our galaxy.

Imagine there is a spacecraft that can take this photo, and you can go out and wave at the exact moment it takes the snapshot. *Three, two, one . . . now!* But the photo will not capture you waving, because the light showing you waving has not yet reached the spaceship. The "now" on a craft that's so far away is not the same as the "now" for you and me on Earth, adding a different meaning to "past," "present," and "future." It all depends.

Cosmic Past, Present, and Future

The night sky lets you look back in time. What is easy to forget is that the same goes for any alien astronomers looking as well. If the Earth were one hundred light-years away from them, they would see it now as it was one hundred years ago. At a distance of five thousand light-years, alien astronomers today would only see the first civilizations bloom on Earth. On a planet one hundred million light-years away, alien astronomers would still see dinosaurs roaming. So if you wonder if beings on other planets know that we—a civilization using technology and inventing space flight—exist here on Earth, that depends on how far away they are.

The Milky Way is about one hundred thousand light-years across, which means that light from a star on the edge of one side needs about one hundred thousand years to cross to a star on the opposite edge. *Homo sapiens* started to migrate from the African continent to parts of Europe and Asia between seventy thousand and one hundred thousand years ago, and that is how a hypothetical astronomer watching Earth from the other side of our galaxy would see our species. There wouldn't be any spacecraft or orbiters exploring the solar system yet. And we would likewise see the home of these curious alien astronomers as it was about one hundred thousand years ago.

Voyager 1, one of the spacecraft carrying the Golden Record we met earlier, was launched in 1977, so the light chronicling the launch has not made it very far. For us, the launch of Voyager 1 is in the past. For an observer on a planet one hundred light-years away, the light that shows Voyager 1's launch has not yet arrived, and therefore Voyager 1's launch is in the future. An observer forty-five light-years away would see the launch at the time I'm writing these words. So did the Voyager 1 launch happen in the past, is it taking place now, or will it happen in the future?

Imagine a grid for our cosmos—spacetime—where space occupies three axes and time is the fourth. If you lie in your bed in the morning and then again in the evening, you are at the same place in space but not the same place in time, so the you in bed in the morning is at a different point in spacetime than the you in bed in the evening. But because the universe itself is embedded in the fabric of spacetime, the *now* and the *past* and the *future* are all simply places in this intriguing cosmic grid relative to any event. Your location in the grid determines what is in the past, present, and future uniquely for you.

The *cosmological principle* states that the universe appears the same everywhere once the scale is sufficiently large. It is the key to figuring out our past because we can't view our past from our own specific vantage point in space time, but we can view the past around us. We are moving along the time axis, always forward. But because light needs time to travel, we get a glimpse into the past of the cosmos elsewhere. Thus, we can look at galaxies farther away from us, which show us only their younger selves, to find out more about what our galaxy was like when it was younger. The history of the cosmos stretches out around us like a beautifully woven tapestry with increasing age toward the cosmic horizon. But there are limits to how far we can see.

The farther away you are from a light bulb, the dimmer it looks. The same goes for stars. Even if you're in a dark place without streetlamps that outshine the stars, you can see only about 4,500 stars in the night sky. The others are just too dim for you to spot. However, you can improve that by using binoculars, which let you see about 100,000 stars. With a three-inch ($\sim$ 8 cm) telescope, you can see roughly 2.5 million stars; with a fifteen-inch ($\sim$ 40 cm) telescope, around 200 million stars. Binoculars and telescopes collect light just like your eyes, and the bigger they are, the more light they catch as we have seen. That is why astronomers build larger and larger telescopes—to catch more and more light from stars that are farther away, seeing further back in time to when the universe was much younger, as we saw in the first JWST deep-field images at the beginning of our story. In images of galaxies around us taken by increasingly powerful telescopes, we can see the cosmos change over time.

But even with the biggest telescopes imaginable, we reach a limit that prevents us from seeing everything: we can only see a fraction of the cosmos. That is because there is a limit to how far light can have already traveled, set by the age of the cosmos. Light did not start traveling before the cosmos existed. So, how old is our universe? Remeber how we figured out how old the Earth is. But there are no meteors that date back to the birth of the cosmos. However, watching the sky revealed the mystery of the age of our universe—and it turns out Earth missed the first two-thirds of it. Just pause for a moment to appreciate that we can see back to a time when neither the Earth nor the Sun, let alone humans, existed.

Edwin Hubble realized in 1929 that the universe expands over time based on the redshift of galaxies as we saw earlier. The Hubble space telescope quite fittingly named after him, made that even clearer. A few years earlier, a Belgian astronomer and priest, Georges Lemaître, had predicted this theoretically. His contribution was recognized by a vote by the International Astronomical Union members in 2018 renaming Hubble's law as the Hubble-Lemaître law. When the Vatican scheduled a celebration of Lemaître's contribution at a scientific meeting of the Pontifical Academy of Sciences in Vatican City in 2016, I received an email asking politely if I would be available to present the latest results on the search for life, alongside other talks by Stephen Hawking, a few Nobel Prize winners, and other esteemed colleagues. And I would meet Pope Francis. Inconceivably, I was able to rearrange my schedule for these fascinating days in Rome.

By measuring how fast other galaxies are moving away from Earth, you can figure out how fast the universe is expanding. Once you know that, you can trace the expansion back in time. If the expansion has been going on at the same rate since its beginning, the universe must have been extremely dense when it started out. Astronomers have figured out that the Big Bang—the birth of the cosmos—occurred about 13.8 billion years ago, so light has not had more than 13.8 billion years to travel. But there is more: at its start, the cosmos was very different from what we see today, which has further refined our calculations about the age of the universe and everything in it. Though the Big Bang was an astonishingly dense and staggeringly hot start of the cosmos, it was not simply an explosion, or at least not the kind we are familiar with, so its name is quite misleading.

Spacetime exploded, but not in just one spot—it exploded everywhere at the same time. That is even weirder than it sounds. The Big Bang, unlike a normal blast, was not an explosion *into* the space around it; there was nothing yet to explode into. It was the birth of space and time itself. After the Big Bang, everything raced away from everything else in an incredibly rapid expansion of the cosmos in a fraction of a second. The Big Bang and the early universe are challenging concepts to imagine, even for astronomers. It means that, initially, every place in the cosmos touched every other place. The farthest galaxies we can spot today far away from each other, were once that close together. The Big Bang was a state of matter and energy so extreme that physics can't even really describe it yet. Cosmologists are working on it, but there is still a lot of unresolved quirkiness.

I find it easiest to imagine this explosion everywhere at the same time as being like observing raisins in raisin-bread dough—all the raisins in the expanding dough move away from one another when the dough rises. But the Big Raisin Bread Expansion is probably not as catchy a name as the Big Bang. At a certain distance from us—at a younger age of the cosmos—stars or galaxies disappear from view: the cosmos had not yet made them. These dark ages of the cosmos ended about 13.5 billion years ago with the first starlight penetrating the darkness. (They are not to be confused with the Dark Ages on Earth. In those Earthly Middle Ages, the sky looked pretty similar to what you see today.)

The speed limit on light, that might sound frustrating, is what allows us this privileged view of most of the universe's past, helping us sketch the evolution of the whole cosmos.

A Baby Picture of the Cosmos

If the past corresponds to a distance from Earth, then if you look far enough away, you should be able to see the incredibly hot and dense era of the early universe in radiation that has traveled for billions of years to get to us. That is precisely what astronomers have found—radiation that shows us a baby picture of the universe. But this picture is distorted because of its long trip. Remember the space-time that the whole cosmos is embedded in? Imagine it as being like a fabric that expands, and this fabric also stretches the light waves that travel through it. So to find the light of an extremely hot, young universe, you need to account for its travel time, and the resulting stretching of the wavelength due to the trip. You need to search for longer, stretched-out wavelengths. That is how scientists found it (rather by chance): it was background noise they could not get rid of no matter where they pointed a microwave antenna. The heat of the early universe is reaching us now from every direction in the sky. Having traveled for billions of years through the cosmos, the ancient light waves that encode the heat signature of the young cosmos were stretched into longer microwaves (although they are not intense enough to heat your food). If we had eyes that were sensitive to microwaves instead of visible light, we would be able to see the leftover radiation from the Big Bang glowing day and night everywhere in the sky, the stunning *cosmic microwave background* (CMB).

The CMB shows that the early universe was extremely hot and that the signal—and thus, the temperature of the sky—is nearly the same everywhere. The differences in the temperature are minuscule—but still, these tiny differences shaped our universe. A change of only 1 part in 100,000 in the density of the young cosmos created everything, because hotter areas are just a bit less dense than cooler ones. For comparison, 1 in 100,000 is about one hour in the life of an eleven-and-a-half-year-old. But these tiny changes added up over time due to gravity. Gravity will pull in material from less dense regions to those that are just a little denser than their neighborhood. This added material makes the gravitational pull even stronger because that region grows in mass compared to its surroundings.

These minute variations in the baby picture of the universe define the structures we see around us today. These ever so slightly denser regions turned into the areas where galaxies formed. With the help of immensely complex computer models, astronomers can trace the evolution of our cosmos from those tiny density differences to where galaxies formed and are located right now. The CMB lets us see the cosmos a mere 380,000 years after the Big Bang, about 13.4 billion years ago. It shows the time when light, which had been trapped in a hot, dense plasma, could finally start to travel through the universe. Before the CMB, the universe was made of this dense, hot plasma. Yes, the cosmos becomes weirder as we go further back in time.

Imagine the CMB as being like a photo of a bright wall of fire, so bright that it keeps us from spotting what is behind it—that is, what came before. This incredibly hot universe—an extremely dense and hot soup of particles—was utterly different from how it is now. At the beginning there were no stars yet; it was way too hot for stars to hold together. Not even atoms were possible; it was too hot for electrons, protons, and neutrons—the pieces you need to form an atom—to survive. But then the cosmos expanded and cooled. Around one-tenth of a millisecond after the Big Bang, the first protons and neutrons formed, then electrons formed as well, and the cosmos continued to expand and cool. Around two minutes after the Big Bang, it was "only" about 1.8 billion °F (~ 1 billion °C), which is much hotter than the center of our Sun but cool enough to let the first atomic nuclei form (protons and neutrons stick together). All of that—from the hot, dense, primordial soup to the first atoms—took about three minutes. Then, for hundreds of thousands of years, the cosmos was basically a hot soup of these nuclei and electrons with light's photons bouncing between them. And the cosmos was expanding and cooling still. About 380,000 years after the Big Bang, the universe had cooled sufficiently for electrons to approach the nuclei and form the first atoms. Light that had been trapped in the plasma, bouncing between the positively charged nuclei and the negatively charged electrons, could for the first time escape.

Since the start, the cosmos changed from an incredibly dense, astonishingly hot plasma to a universe where light started traveling freely—with its temperature painting the CMB on our night sky. About 100 million years later, the first star made of those ancient elements ignited, setting the stage for the Sun, Earth, and curious humans. So, even though we have not found an end or a rim of the universe, the area of the universe we can observe has a hard limit. Wherever in the cosmos you are located, if you have large telescopes to look at ever-dimmer objects ever farther away, you will finally arrive at this picture of the hot plasma of the early universe. When ancient mapmakers ran up to the very edge of the known world, they wrote *Hic sunt dracones*, or "Here be dragons," the medieval practice of putting illustrations of dragons, sea monsters, and other mythological creatures on uncharted areas of maps where danger could lurk. More precise for cosmic maps is "Here be astonishingly hot plasma." It limits any cosmic exploration—a time horizon that shapes a sphere around us of the *observable universe*. That means our view is limited by the ever-expanding distance from which light has had time to reach us since the beginning of the cosmos.

But those limits allow us to explore a vast region of space-time around us. From the first stars and galaxies sparkling in the JWST deep-field image to the thousands of stars with intriguing exoplanets on our cosmic shore. We live in an era of fascinating cosmic exploration.

Every point in space is at the center of its own sphere of its observable universe, rimmed by the hot plasma of the young cosmos. And wherever you are looking at it— from Earth or from a galaxy far, far away—you are in the center of your own observable universe. And everyone else, including any aliens elsewhere in the distant past or future, are also at the center of *their* observable universe.

So you could argue that humans have regained our special place at the center of the (observable) universe.

Who Could Be Watching Us Right Now?

If somewhere in the cosmos life exists, and if these entities are curious, like we are, could they be watching us? We stand on the threshold of finding life in the cosmos. We have found more than five thousand exoplanets, and thousands more will be added to our list soon after the initial tests on these signals are concluded. With the JWST launched, we have a big enough telescope to collect light from close-by exoplanets that could be like ours. If other civilizations exist and have only *our* level of technology, could they spot us?

When astronomers catalog the stars in our solar neighborhood, looking out to a mere three hundred light-years away, they find about three hundred thousand objects— mostly stars and stellar corpses. Most of them are cool red stars, and all move in a mesmerizing gravitational dance around the center of our galaxy, the Milky Way. In our search for other worlds we focus on planets that block part of their star's light from our view: transiting worlds. But for that to happen, the alignment of the planet, its star, and us must be just right. So, there is also a Goldilocks zone for observing planetary transits.

While trying to find alien Earths I started to wonder which stars are in the right place to spot us? Which stars have a cosmic front-row seat to witness Earth block out some of the light from our Sun? Where are *we* the aliens?

As it turns out, together with the inspiring American astronomer and curator at the American Museum of Natural History, Jackie Faherty, I figured out which stars could spot us. We took the opportunity provided by the Gaia mission, designed to generated a precise star catalog of our solar neighborhood, containing not only each star's position but also how it moves. This allowed us to map the nearby stretch of our galaxy in the spacetime coordinates. Combing through the Gaia database, we identified the stars that were just at the right place to see the Sun dim ever so slightly because Earth temporarily blocked part of *their* view of our host star.

Fewer than one out of a hundred of our neighboring stars can see that slight dimming because Earth is small: those stars with a position close to the ecliptic, the plane of Earth's path around the Sun we encountered earlier. About fifteen hundred star systems would be able to see the slight dimming of our Sun due to the Earth this year. That number increases to nearly two thousand if you move the dial of time back and forth and expand the window to anyone looking from about five thousand years ago to five thousand years into the future.

About a hundred of these stars are so close to Earth that our radio waves have already swept over them. On these worlds, curious alien astronomers would be able to not only spot us but also listen to our somewhat eclectic taste of music. We have already spotted that three of those systems host exoplanets in their habitable zones. Any civilization with our level of technology on Ross 128 b, for example, which circles a red dwarf star in the constellation Virgo a mere eleven light-years away, might already have seen and heard us. But they cannot see us anymore because their perfect vantage point started about three thousand years ago and ended about nine hundred years ago. Would alien observers have concluded that there was intelligent life on Earth nine hundred years ago? Observers on a planet circling Teegarden's Star, about 12.5 light-years away, will start to see the Sun ever so slightly dim in 2050 but they might have heard us already. The fascinating TRAPPIST-1 system that we have encountered earlier, at only forty light-years away, will be able to see the Sun dim starting about sixteen hundred years from now.

As these examples show, this special vantage point is not guaranteed forever. It is gained and lost in the precise gravitational dance in our dynamic cosmos. And it is not a case of "If I can see you, you can see me." It is more like two ships passing in the night—some you can see, while some others can see you. So how long does that cosmic front-row seat, where an observer can see the Earth block the light from the Sun, generally last? To answer that, we dove into the Gaia data on the motion of the stars, which allowed us to track the movement of the stars into the future and trace it back into the past. We found that this vantage point lasts for at least a thousand years, so if we had searched the sky for transiting planets thousands of years earlier or later, we would see different ones. And different ones could spot us.

Whenever I look up at the sky now, I imagine the two thousand close-by star systems that could see our planet block just the tiniest bit of the light from our Sun. If there is intelligent life out there that has already found us, I like to imagine that it is rooting for us. It might be starring in a cosmic reality-TV show with different episodes: *Oh no, look, they are generating an ozone hole!* And then *Oh, look, they fixed it—go, planet!* Followed by *Oh no, now they are changing the planet's climate.* I hope there will be an *Oh, look, they are fixing the climate—go, planet!* coming soon.

If someone has found us already, I wonder what they think of us.

Spaceship *Earth*

In the far future, imagine you are undertaking the journey to one of these new worlds upon which we have found clear evidence of life—our first alien Earth. To make such a trip you will need a ship only dreamed of so far in science-fiction stories, but one day we might build it. These working starships will need to carry everything you require to survive and be able to recycle it perfectly for the long voyage ahead. The integrity of your ship is your protection against an incredible hostile environment: space.

Preparing for launch, you go over your pre-flight checklist to give you the best chances of survival on your journey through unforgiving, unexplored terrain. You survey the spacecraft subsystems (food, water, air, propulsion, navigation) to make sure all parts function optimally. Your water tanks integrate into the hull of your ship for added radiation protection. The spaceship is a marvel of connectivity that creates the perfect biosphere with the right amount of oxygen, about 21 percent. You could live with a bit less, but this way it is easy to breathe even when you walk fast.

You test the water for any contamination. Everything needs to work perfectly on this ship to not disturb the delicate balance in the biosphere that keeps you and your fellow travelers alive. The large vertical hydroponic bays and the gardens covering the walls of many compartments produce food and filter out CO_2. The amount of biota in the soil and the water generates just the right mix of chemicals for you to breathe. Next, you check the crops to ensure you'll have fresh food to survive. The hydroponic tanks hum comfortingly, and the soil composition looks promising for a good harvest and for planting new seeds. You have stored parts of the earlier crop, but fresh food will help provide the vitamins that keep you healthy, a critical part of your diet. You breathe in the smell of wet soil as your steps echo on the bridge of your spaceship while you finish your inspection.

After takeoff, you look out the window at Earth, and you wonder why it took leaving to recognize that this incredible planet is also a ginormous spacecraft, covered by a dazzlingly effective life-support system, the biosphere, a complex of balanced networks that keep us and so many other species alive.

Only now do you realize what you are leaving is Spaceship *Earth*, our home on our long journey through the cosmos. Its fate inextricably bound to that of the solar system. A spaceship we need to care for better, even as we begin to journey beyond. I like to picture us as knowledgeable stewards, learning to protect the only home we have ever known.

I like to envision humanity's future living in an unspoiled environment, because we have found ways to robotically process resources in space, where poisonous gases do not contaminate the air we breathe and the water we drink. There is no planet that will ever be as perfectly suited for us as Earth is—from microbes to humans, we evolved together with our astonishing Pale Blue Dot.

Exploring space allows us to gather the knowledge to save ourselves from asteroids, from pollution, and from using up the limited resources on Earth, our beautiful, incredibly complex yet fragile "mote of dust suspended in a sunbeam," as Carl Sagan so eloquently described it.

And yet, on this mote of dust we are already creating the first travel charts for future interstellar explorers. We are putting intriguing destinations on the map, from stunning lava worlds to places where you can chase two shadows. We do not have the ships to venture into the cosmos yet, but we have found other ways to explore the universe deciphering the messages encoded in light. The planets in our solar system have revealed profound lessons on the transforming yet sometimes fragile concept of a habitable world. Combined with the stunning diversity of life on our Pale Blue Dot striving even under extreme conditions, adapting to and altering our world throughout history, these insights allow us a glimpse of the first planets that could be alien Earths on our cosmic shore.

Even though we cannot set foot on these new worlds yet, our exploration has changed our view of the sky forever. The thousands of other suns you can see on a clear night hold a breathtaking promise that we just might find someone else out there.

Look up into the stunning sky, the window connecting us to the cosmos. Find your favorite star and allow yourself the freedom to wonder.

What if we are not alone in the cosmos?

Acknowledgments

This book is written from my point of view, but the work described has been undertaken with colleagues from all over the world—in international, interdisciplinary teams I am delighted and honored to be part of. I wish I could have included all your research and names, but I had a set page limit and way too many fascinating discoveries to squeeze in. I also want to thank my colleagues who have been and are trailblazing in this exciting frontier of finding life in the cosmos. I wish I could have mentioned you all. The search for life in the cosmos would not progress without your breakthroughs!

There would be no Carl Sagan Institute at Cornell without the inspiring Ann Druyan. Thank you so much for believing in our idea and your unwavering love for the cosmos. I am grateful to the interdisciplinary team at the Carl Sagan Institute for filling my life with intriguing scientific queries, exciting discoveries, camaraderie, and so many new ideas it is no wonder the espresso machines stay busy trying to keep up. And my incredible research team members—current and past—I love coming to work every day because I know I will meet you there and our discussions will be creative, insightful, and inspiring, leaving me ever grateful that you chose me as your adviser, colleague, and partner in crime for part of your way. A special thank-you to Ligia, Becca, and Jonas for the time they took away from colorful biota, planets on the edge, and lava worlds to provide insightful comments and positive feedback on the manuscript.

I am deeply grateful to Ann, Bill, Charles, David, Eric, Laetitia, Linda, Mike, Peter, Sam, Scott, Shami, Steve W., and Steve Z. for their generosity to read and comment on the manuscript—your help allowed *Alien Earths* to shine (and hopefully has eliminated most of my errors).

I was overwhelmed by the kindness of other scientists and authors in giving advice; my deep gratitude to Alan Alda, Charles Cockell, Marcelo Gleiser, Chris Hadfield, Robert Hazen, Kathie Mack, Martin Rees, Caleb Scharf, Steve Strogatz, and Neil deGrasse Tyson. And to Alastair Reynolds and Andy Weir, I really enjoy our always inspiring discussions of sci-fi worlds and if they could exist in real life. And thank you for all the advice that you generously provided.

There are so many people in my life who have provided friendship, mentorship, support, and advice for which I am deeply grateful. I chose not to name you here, but please know that you have made and continue to make a huge positive impact in my life. I am trying to support and encourage others modeling how I do that after how you have done that for me.

I would have never made it where I am today without my incredible family, foremost my parents and my sister, who always believed in me, but also my wider family members, who are always excited to hear about the newest discoveries and forgive me for sometimes being too busy to call more often to catch up. My family and my incredible network of friends all over the globe make our Pale Blue Dot my favorite planet in the cosmos. And this book is dedicated with love to Filipe and Lara Sky, who make every day a beautiful new adventure.

I want to thank my wonderful agent, Deirdre Mullane of Mullane Literary Associates, and my amazing editors Elisabeth Dyssegaard and Daniela Rapp, who fell in love with my book when it was just an idea, and to Jamilah Lewis-Horton and the fantastic team at St. Martin's Press. Thank you for taking a chance on the idea of *Alien Earths*. I am grateful to Keith Mansfield for his unwavering enthusiasm and warm, sage advice and the team at Penguin Books for seeing my book for all it could be. Thank you, Tracy Roe, for the sparkles when I needed them most, and it was just a delight to watch the incredibly talented Peyton Stark bring my vision to life in her gorgeous art that you find throughout the book (https://www.peytonstarkstudio.com/).

I wrote this book mostly during my sabbatical in Salzburg and Lisbon, where I could steal away a few hours a day to write, and I am grateful to the Paris Lodron University of Salzburg and to the Instituto Superior Técnico,

Universidade de Lisboa for providing me a place to belong and a quiet office to think and write. I am grateful to the Brinson Foundation for believing in my research and in this book. A special thank-you to the Simons Foundation, the Heising-Simons Foundation, the Kavli Foundation, and the Breakthrough Foundation for supporting my team's quest to find life in the cosmos. And every writer needs a coffee shop in which to write—here is a special thank-you to the team at the Insensato Café-Livraria in Tomar, Portugal, who never minded that I was typing away on my laptop for hours with a Portuguese espresso or a fresh tea steaming in front of me to fuel my thoughts.

A warm thank-you to everyone who ever said something positive to me. You made searching for life in the cosmos just a little bit easier.

And my last thank-you is to *you*, for picking up this book and following me on the adventure to find life in the cosmos. I hope it has changed your view of the night sky too. We are living in an incredible time of exploration. I am rooting for us!

The Golden
Record Playlist

1. Greeting from Kurt Waldheim, Secretary-General of the United Nations 00:43
2. Greetings in 55 Languages 03:46
3. United Nations Greetings/Whale Songs 04:04
4. Sounds of Earth 12:18
5. Munich Bach Orchestra/Karl Richter - Brandenburg Concerto No. 2 in F Major, BWV 1047: I. Allegro (Johann Sebastian Bach) 04:43
6. Pura Paku Alaman Palace Orchestra/K.R.T. Wasitodipuro - Ketawang: Puspåwårnå (Kinds of Flowers) 04:46
7. Mahi musicians of Benin - Cengunmé 02:10
8. Mbuti of the Ituri Rainforest - Alima Song 01:00
9. Tom Djawa, Mudpo, and Waliparu - Barnumbirr (Morning Star) and Moikoi Song 01:29
10. Antonio Maciel and Los Aguilillas with Mariachi México de Pepe Villa/Rafael Carrión - El Cascabel (Lorenzo Barcelata) 03:19
11. Chuck Berry - Johnny B. Goode 02:40
12. Pranis Pandang and Kumbui of the Nyaura Clan - Mariua-mangi 01:24

13. Goro Yamaguchi - Sokaku-Reibo (Depicting the Cranes in Their Nest) 05:04

14. Arthur Grumiaux - Partita for Violin Solo No. 3 in E Major, BWV 1006: III. Gavotte en Rondeau (Johann Sebastian Bach) 02:57

15. Bavarian State Opera Orchestra and Chorus/Wolfgang Sawallisch - The Magic Flute (Die Zauberflöte), K. 620, Act II: Hell's Vengeance Boils in My Heart (Wolfgang Amadeus Mozart) 02:59

16. Georgian State Merited Ensemble of Folk Song and Dance/ Anzor Kavsadze - Chakrulo 02:20

17. Musicians from Ancash - Roncadoras and Drums 00:54

18. Louis Armstrong and His Hot Seven - Melancholy Blues (Marty Bloom/Walter Melrose) 03:06

19. Kamil Jalilov - Muğam 02:34

20. Columbia Symphony Orchestra/Igor Stravinsky - The Rite of Spring (Le Sacre du Printemps), Part II—The Sacrifice: VI. Sacrificial Dance (The Chosen One) (Igor Stravinsky) 04:38

21. Glenn Gould - The Well-Tempered Clavier, Book II: Prelude & Fugue No. 1 in C Major, BWV 870 (Johann Sebastian Bach) 04:51

22. Philharmonia Orchestra/Otto Klemperer - Symphony No. 5 in C Minor, Opus 67: I. Allegro Con Brio (Ludwig van Beethoven) 08:49

23. Valya Balkanska - Izlel e Delyu Haydutin 05:03

24. Ambrose Roan Horse, Chester Roan, and Tom Roan - Navajo Night Chant, Yeibichai Dance 01:00

25. Early Music Consort of London/David Munrow - The Fairie Round (Anthony Holborne) 01:19

26. Maniasinimae and Taumaetarau Chieftain Tribe of Oloha and Palasu'u Village Community - Naranaratana Kookokoo (The Cry of the Megapode Bird) 01:15

27. Young girl of Huancavelica - Wedding Song 00:41

28. Guan Pinghu - Liu Shui (Flowing Streams) 07:36

29. Kesarbai Kerkar - Bhairavi: Jaat Kahan Ho 03:34

30. Blind Willie Johnson - Dark Was the Night, Cold Was the Ground 03:21
31. Budapest String Quartet - String Quartet No. 13 in B-flat Major, Opus 130: V. Cavatina (Ludwig van Beethoven) 06:41

Some of the recording names and performers have been updated from the original listing due to initial errors in the material provided to the Voyager Interstellar Record team. David Pescovitz and Tim Daly from Ozma Records reissued the Golden Record—for the first time on vinyl—for Earthlings to enjoy in 2017. Through in-depth research they identified and corrected errors and omissions in the original information—and won a Grammy in 2018. Jonathan Scott wrote about the detective search for the title and musicians of some of these amazing tracks in his cheerful 2020 tour guide, *The Vinyl Frontier*.

The book *Murmurs of Earth* describes this incredible record and was written in 1978 by the six people who created it: Carl Sagan (chair of the committee), Frank Drake (technical director), Ann Druyan (creative director), Timothy Ferris (producer), Jon Lomberg (designer), and Linda Salzman Sagan (artist). They address the daunting challenge of deciding which songs, pictures, and sounds to include on the record that would become a time capsule of the story of our world—a gift from a Pale Blue Dot to the cosmos.

To Learn More

Where are the Golden Records now? Follow the Adventure of Voyager 1 and Voyager 2 and the route of the Golden Records through space: https://voyager.jpl.nasa.gov/mission/status/ or follow @NASAVoyager.

Why are the numbers rounded off? You might wonder why the numbers you encounter in this book are rounded off, not the exact values. As authors Chip Heath and Klara Starr suggest in their book, *Making Numbers Count*, rounding numbers and providing a comparison to things we all know is a useful way to translate data into "stories that stick." As I share the fascinating beauty of the cosmos with you, I tried to make its weird and wonderful characteristics stick, perhaps losing a few less important digits but keeping the overall grandeur of scale intact.

How to send your name to Mars: NASA provides a special opportunity: you can send your name to destinations throughout our solar system (and can get boarding passes for any friend you like, which makes a great birthday present), for instance. For the next flight to Mars (https://mars.nasa.gov/participate/send-your-name/mars2020/). Maybe some future alien archeologist will try to decipher your name when exploring the solar system?

Want to name an exoplanet? Anyone can propose names for exoplanets to the International Astronomical Society via the NameExoWorlds campaign started in 1995. See details at https://www.nameexoworlds.iau.org

Want to help find new worlds or solve other science mysteries? Citizen science projects are collaborations between scientists and interested members of the public, open to anyone around the world. See: https://science.nasa.gov/citizenscience. Volunteers, known as citizen scientists, have helped make thousands of important scientific discoveries through these collaborations, and have found new worlds, like in the case of Kepler-64 b or PH1 b (Planet Hunters 1).

How to get NASA's vintage travel posters: NASA's free vintage travel posters are titled Visions of the Future. Colorful and inventive images of travel destination in our solar system and beyond instill both optimism and a desire for a future where humanity is traveling between the stars. The exoplanet section is called Exoplanet Travel Bureau. Inspired by the vintage travel posters we also created for our research at the Carl Sagan Institute (see https://www.jpl.nasa.gov/galleries/visions-of-the-future).

Update on new discoveries of the Carl Sagan Institute: The Carl Sagan Institute was founded by Lisa Kaltenegger in 2015 to find life in the cosmos. Based on the pioneering work of Carl Sagan at Cornell University, our interdisciplinary team with scientists from fifteen different academic departments is developing the forensic toolkit for our search both inside the solar system and on planets and moons orbiting other stars. Want to know more and stay in touch?

Webpage: https://carlsaganinstitute.cornell.edu/
Instagram: https://www.instagram.com/carlsagani/
Twitter: @CSInst
YouTube: https://www.youtube.com/c/CarlSaganInstitute

Index